HUDSON VALLEY WARMING

Jerome S. Thaler

ACKNOWLEDGEMENTS

A book of this kind would not be possible without the daily input of weather observations by dedicated observers of the National Weather Service Cooperative network and others who have taken readings and made measurements for many years. Other contributors have been the Hudson Valley Climate Service for its file of extremes. David Lawter for the book layout and Eleanor Arnold for editing.

AUTHOR'S PUBLICATIONS

The Westchester Weather Book, 1977, George Candreva Environmental Center, Yorktown Heights, NY

Putnam County, NY-Climate Summary, Rotary Club of Lake Mahopac, NY

West Point, 152 Years of Weather Records, 1979, Weatherwise, 32, 112-115

Daily Extremes of Temperature, Rain and Snow for Selected Long Term Hudson Valley Weather Stations, Unpublished

A Cyclical Pattern in 19th and 20th Century Hudson Valley Winter Temperatures, 1983 Weather and Climate Responses to Solar Variations, 189 – 196 Ed. BM McCormac et al., Colorado Associated University Press

Hudson Valley Reconstructed Temperature Sets: History and Spectral Analysis, 1987 Climate – History, Periodicity and Predictability, 70 – 77, Ed. MR Rampino et al. Van Nostrand R-einhold Co.

Hudson Basin Precipitation Regimes, Croton, Esopus and Sch.oharie Watersheds, 1989, Northeastern Environmental Science, 106 – 118, 8, 2

Climate of.Albany County, 1994 Northeastern Geology and Environmental Sciences, 162 – 193, 16, 3&4

Catskill Weather, 1996, Purple Mountain Press, Fleischmanns, NY

Hudson Valley Seasonal and Annual Mean Temperatures: History, Trends and Periodicities, 1998, Northeastern Geology and Environmental Sciences, 133 – 144, 20, 2

Weather History and Climate Guide to the Lower Hudson Valley, 2001, George Candreva Environmental Center, 820 Pinesbridge Road Ossining., NY

Adirondack Weather, 2004, Hudson Valley Climate Service, 1720 Morningview Drive, Yorktown Heights, NY 10598

Articles and data contributions on local weather have appeared since 1964 in the Peekskill Star, Reporter Dispatch, New York Times - Westchester Edition, Patent Trader, Yorktowner, North County News, Peekskill Herald, Joumal News - newspapers and in Hudson Valley, Adirondack Life and Catskill Life - magazines.

Presentations on local climate have been presented to the general public at least yearly for over the last decade.

CONTENTS

INTRODUCTION

At the present time global warming and its probable causes have been the dominant environmental issue worldwide. There is scarcely a media outlet (books, newspapers, radio, periodicals, television, and internet) that has not pointed out this fact. Most recently climatologists have noted that the warming is already on us and is reaching the point that little will be able to be done before the serious adverse effects on all life will occur.

This study is not concerned about weather, which is what is happening at any given time, but about aspects of climate that are based on data and statistics built up over many years.

Most climate studies since 2000 have shown that there has been significant world wide warming since the late 20th century. While the amount of warming revealed differs in degree from one continent to another there has been no plausible scientific evidence to the contrary. Polar regions particularly, have the greatest warming as indicated by ice melting and recession as well as by decades of temperature measurements. In addition the precipitation regimes in many areas have been affected. Within the Hudson Valley watershed region the speed of the warming is indicated by the frequency of new records being set in the period since 2000. Warmest years (in the top ten) are indicated as well as the number of warmest winters, warmest seasons and warmest months since 2000.

The purpose of this study is to focus on one region and isolate the numerous elements that reveal the timing and amount of the warming that has occurred. For that purpose the Hudson Valley watershed is ideally suited for such a study. About three million people reside in the counties north of New York City that make up the watershed. Weather station locations by county are Westchester, Dutchess, Columbia, Rensselaer, Orange, Ulster, Albany, Saratoga, Warren, Oneida, Essex, and Hamilton. Within these are eighteen weather stations that have recorded and maintained continuous daily temperature observations for fifty or more years. Nearly all of these are part of the National Weather Service network of cooperative stations that measure temperature, precipitation and other weather phenomena. Other reporting stations not included in this study have fewer or missing years.

Data from these stations has been published monthly since the 1880s in the following government publications: Monthly Weather Review 1887 to 1912; Department of Agriculture Weather Bureau New York Section 1913 to 1939; Department of Commerce Weather Bureau New York 1940 to 1975; National Oceanic and Atmospheric Administration publication: Climatological Data, New York 1976 to 2010. . This data is now available on the internet from NOAA. Older data from the mid 1800s can be obtained on microfilm.

While there are other weather stations in all of the counties of the Hudson Valley watershed, not all of them are of National Weather Service quality in terms of instrumentation nor have they been in operation for at least 30 continuous years.

For residents within the Hudson Valley watershed, there is a need to know the extent of the warming in their local community and how the overall temperature rise has impacted other temperature related phenomena plus that of precipitation.

The U.S. Weather Service Cooperative stations that provide the basic data are in many cases located in or near the cities and towns in their county.

Within the Hudson Valley climate division the cities nearest to stations are as follows: In Westchester county White Plains – Westchester Airport; Peekskill – Peekskill Water Dept; Yorktown Heights – Yorktown US coop station; In Orange county Newburgh – West Point US coop station; Middletown – Middletown US coop station; In Dutchess county Poughkeepsie – Poughkeepsie US coop station; Millbrook – Millbrook station; In Ulster county – Mohonk Lake US coop station near New Paltz and Slide Mountain -US coop station near Phoenicia; In Albany county - Alcove Dam,& Albany Airport US coop stations;   In Rensselaer county at Troy – Troy Lock & Dam US coop station; In Saratoga county -Saratoga Springs – Saratoga Springs coop station; In Warren county -Glens Falls – Glens Falls Farm US coop station, and Glens Falls Airport.

Within the Mohawk Valley climate division are three weather stations as follows: In Oneida county -Gloversville – Gloversville US coop station; Little Falls – Little Falls City Reservoir US coop station; and Utica – Utica Airport US coop station.

Located within the Northern Plateau climate division of the watershed are the Indian Lake US coop station in Hamilton county and the Newcomb US coop station in Essex county.

Within the last five years or more some of these station have become inactive or have had broken or missing months with measurements not taken or not published. These are Middletown, Slide Mountain, Glens Falls Farm, Gloversville, and Utica.

CLIMATE DIVISION STATION LOCATIONS

Long term weather stations maintained by the National Weather Service as part of its cooperative observer network are the data source and basis of this study. These stations are located in the climate divisions that make up all or part of the Hudson Valley watershed (see Figure 1) namely - Hudson Valley, Eastern Plateau, Mohawk Valley, Northern Plateau and the Coastal division. The main network of sixteen stations used in this work have all taken daily weather readings from 60 to 180 years. Within the Hudson Valley climate division there are eleven reporting stations and one with unpublished data; there is one in the Eastern Plateau division; there are three in the Mohawk Valley division; and two in the Northern Plateau division. Two other stations are located in the northern fringe of the Coastal division.

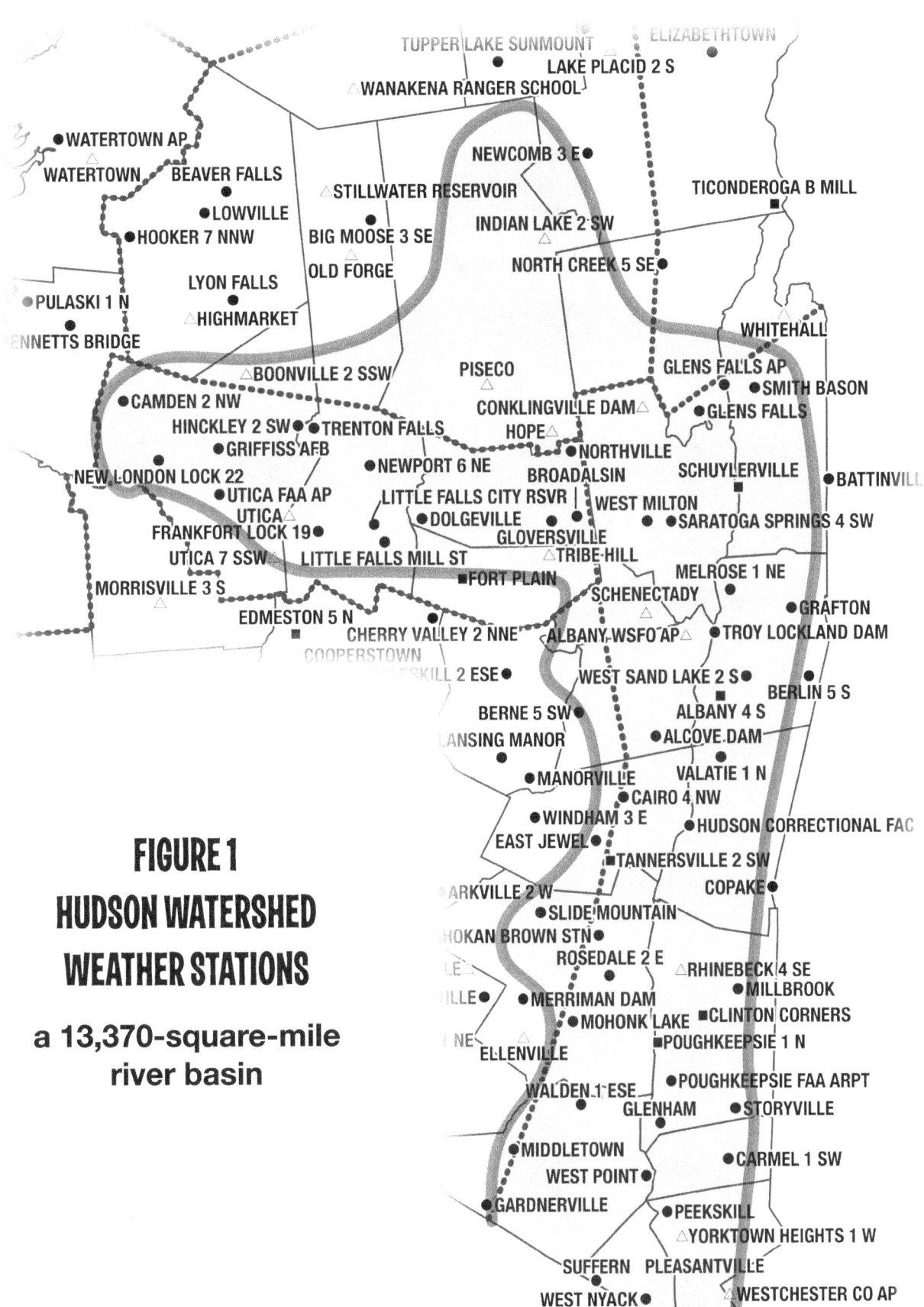

FIGURE 1
HUDSON WATERSHED
WEATHER STATIONS

a 13,370-square-mile
river basin

TIMELINES AND FREQUENCIES OF HUDSON VALLEY WARMING

The determination of the warming of the Hudson Valley watershed region is based on the arrangement of data from the network of weather stations some of which have been in operation before the beginning of the 20th century. The following are aspects that show the trends of temperature and frequencies of heat.

Warmest years, seasons, months and days

Climate Division temperature trends over 30 and 10 year periods

Heating Degree trends

First and last frost of season

Frost free season length and trend

Frequency of extreme heat and cold

Hardiness Zone changes

Number of seasonal days with daily minimums equal or greater than 70 degrees.

Hottest daily extreme temperatures

Ice on – ice out dates

Length of Ice on

Greatest Annual Precipitation

Wettest months and years

Greatest daily precipitation.

TIME FRAME WARMINGS

The warming that has characterized the Hudson Valley watershed in the 20th century and continuing through 2012 can be revealed by using different time periods. The National Weather Service uses 30 year averages of all reporting stations for determining what is normal for temperature and precipitation. As of 2011 the 30 year average is the period of 1981 – 2010 Thus an examination of past 30 year periods reveals how our current average compares with those of the past. The climate divisions of the watershed show (Tables 1,2,3) that the coldest period for all divisions was the first 30 years of the 20th century. Since then the Northern Plateau Division which has some stations outside the watershed shows the 1971 – 2000 period as its next coldest. All divisions have the 1981 – 2010 period as their warmest. Ten individual stations also show (Table 4) the current 30 year period as their warmest.

When 10 year averages (decadal) are employed all climate divisions show that their coldest decades have occurred in the early part of the 20th century (Table 5). The warmest decade for two divisions was the 1951 – 1960 period that surpassed the current one by less than the .5 Degree F. The Hudson Valley division for the 1981- 2010 ranked third in warmth. In (Table 6) ten of eleven stations show their warmest decade in the most recent ten year decade.

Annual averages for 16 stations show that about half of their warmest ten years have occurred in the 2000 – 2011 period (Table 8).

When individual years are compared (Table 7) they show that the warmest year for the Hudson Valley division was 1998; for the Mohawk Valley was 1953 and for the Northern Plateau was 1998. See (Table 8) for the 10 warmest years for each station.

The warmest winters for eleven active stations are the 2001 – 2002 and 2011 – 2012 seasons. The warmest summer season for ten of fifteen stations was 2005.

Since the winter of 2010 - 2011 not only have all monthly averages been higher than the 1981-2010 normal average, but a six month period since November 2011 has had averages in the top warmest ten for the period of record of nearly all watershed stations (Table 9). Ten of fifteen stations have 2005 as their warmest summer, with many since 2000 (Table 10).

MEASURING TEMPERATURE

Before the arrival of the thermometer (invented in the early 1700s) into the Americas, the determination of temperature was a very subjective judgment. Extreme heat was noted for its effect on crops, livestock and people. Coldness was noted in diaries when bodies of water froze over or when the first or last frost appeared. These recorded observations, while useful could not give any objective comparison of climate of one year to the next or even of one month of one year to the same month of a different year.

The first continuous instrumental temperature observations were taken at a few of the oldest educational institutions located in the populated centers of the original American colonies. Harvard and Yale were recording daily temperatures from the mid 1700s thus providing some data forming the basis of the climatic norms and variability of that time. After the American Revolution nearly all major cities had thermometers.

Since the late 1800s, the standard for determining the mean temperature has been adding the high and low temperature for the day and dividing the sum by two. Before then, temperature readings were taken from one thermometer a few times each day, but an exact maximum and minimum could not be determined. The invention of the minimum/maximum thermometer enabled the recording of an exact daily mean temperature. The sum of the daily means divided by the number of days of the month gives one the mean or average temperature for the month. Monthly means are also affected by the daily time of observation. A correction factor has to be employed making midnight the observation time.

TABLE 1
HUDSON VALLEY DIVISION
30 Year Temperature Averages

	Jan	Feb	Mar	Apr	May	Jun	Jul	Aug	Sep	Oct	Nov	Dec	Ann
1901-30													47.5
1931-60	24.7	25.5	34.3	46.8	58.1	67.1	71.9	69.8	61.9	51.4	40.0	28.0	48.3
1941-70	23.3	25.2	34.5	47.0	57.5	66.9	71.5	69.3	61.9	51.7	40.2	27.4	48.0
1951-80	22.9	24.9	34.6	46.8	57.4	66.5	71.2	69.3	61.5	50.7	40.0	27.9	47.8
1961-90	22.1	24.7	35.0	46.4	57.6	66.2	71.2	69.1	61.3	50.2	40.0	27.8	47.6
1971-00	23.1	25.8	35.4	46.6	57.9	66.2	71.1	69.1	61.0	49.7	39.8	28.8	47.9
1981-10	23.4	26.7	35.1	47.3	57.7	66.6	71.2	69.5	61.6	49.9	40.3	29.1	48.2

TABLE 2
MOHAWK VALLEY DIVISION
30 Year Temperature Averages

	Jan	Feb	Mar	Apr	May	Jun	Jul	Aug	Sep	Oct	Nov	Dec	Ann
1901-30													44.4
1931-60	20.1	20.6	29.6	43.2	55.2	64.9	69.5	67.4	59.6	48.7	36.8	23.8	45.0
1941-70	19.0	20.6	30.2	43.7	54.7	64.5	69.0	66.8	59.6	49.3	37.3	23.6	44.9
1951-80	18.8	20.5	30.4	43.5	54.9	64.2	68.7	66.9	59.3	48.4	37.2	24.3	44.8
1961-90	18.6	20.7	31.1	43.5	55.3	63.9	68.9	66.8	59.0	48.1	37.4	24.2	44.8
1971-90	19.5	21.9	31.1	43.5	55.9	64.2	69.2	67.1	59.0	47.6	37.4	25.7	45.2
1981-10	20.0	22.9	31.0	44.5	55.5	64.5	69.0	67.3	59.6	47.7	38.2	26.3	45.5

TABLE 3
NORTHERN PLATEAU DIVISION
30 Year temperature Averages

	Jan	Feb	Mar	Apr	May	Jun	Jul	Aug	Sep	Oct	Nov	Dec	Ann
1931-60	15.0	14.5	28.3	40.1	55.2	61.4	61.9	62.9	53.5	44.9	29.5	18.4	40.5
1941-70	16.4	17.0	25.8	39.6	52.3	61.4	65.7	63.7	56.3	45.8	33.5	20.0	41.5
1951-80	14.9	16.4	26.6	39.5	51.9	61.0	65.1	63.2	56.0	45.3	33.8	20.0	41.0
1961-90	14.3	16.3	27.0	39.3	52.0	60.5	65.1	63.1	55.5	45.0	33.6	20.0	41.0
1971-00	14.7	17.0	26.5	39.0	51.7	60.1	64.7	62.7	54.6	43.5	33.0	20.8	40.7
1981-10	15.6	18.3	26.7	40.2	51.8	60.7	64.9	63.3	55.8	44.0	33.9	21.7	41.4

An examination of the 30 year running averages for the three Hudson Valley climate divisions shows that for one of the three, the most recent period is the warmest. The Hudson valley division (Table 1) for this period is nearly tied in rank with the 1931 – 1960 period for its warmest. The Northern Plateau shows the 1941 – 1970 period also slightly higher than the most recent period.

Ten year averages for each of the climate divisions show the decade of 2001 – 2010 as the warmest of the 20th century for the Northern Plateau with the 1951 – 1960 decade the warmest for the other two divisions. The most recent decade places second.

TABLE 4
30 YEAR STATION ANNUAL AVERAGES

	Albany Airport	Mohonk Lake	Peekskill	Poughkeepsie	Saratoga Springs
1931-60	47.6	47.0	48.1		
1941-70	47.4	46.8	48.6		
1951-80	47.2	46.7	49.1		
1961-90	47.3	47.1	50.0	49.1	47.0
1971-00	47.5	47.9	50.6	50.4	47.1
1981-10	48.2	49.9	51.2	50.1	48.1

	Troy Lock & Dam	West Point	Carmel/ Yorktown	Indian Lake	Millbrook	Little Falls CR
1931-60		50.0	49.3	40.5		44.7
1941-70		48.8	49.1	40.3		44.7
1951-80		49.6	48.8	40.1	47.5	45.3
1961-90	48.0	49.4	49.1	39.9	47.4	46.3e
1971-00	48.8	49.5	50.0	40.3	47.6	44.4
1981-10	49.2	50.5	50.6	40.6	48.2	44.9

e = estimated

An examination of 30 year annual temperature averages for eleven representative long-term Hudson valley weather stations reveals that the most recent period has been the warmest for all stations. The National Weather Service uses the most recent period as its normal for annual, monthly and daily temperatures and precipitation.

TABLE 5
DIVISIONAL DECADAL TEMPERATURE AVERAGES

	Hudson Valley	Mohawk Valley	Northern Plateau
1901-10	47.4	44.6	41.1
1911-20	47.3	44.4	40.1
1921-30	47.8	44.3	41.3
1931-40	48.8	45.5	41.0
1941-50	49.0	45.7	41.6
1951-60	49.2	46.1	41.6
1961-70	47.8	45.1	41.1
1971-80	47.9	45.0	40.6
1981-90	48.3	45.6	41.4
1991-00	48.4	45.6	41.3
2001-10	48.8	46.0	42.3

TABLE 6
ANNUAL DECADAL AVERAGE TEMPERATURES

	Albany	Glens Falls	Mohonk Lake	Poughkeepsie	Saratoga Springs
1901-10	46.5	45.5	45.2		
1911-20					
1921-30	47.8	44.2	46.1		
1931-40	47.8		46.7	49.5	
1941-50	47.6		47.4	49.9	45.5
1951-60	48.0	46.3	47.3	50.2	
1961-70	46.8	44.6	46.5	49.9	44.7
1971-80	46.9	44.7	46.6	49.4	44.6
1981-90	47.8	44.8	47.7	48.9	46.3
1991-00	48.1	45.8	48.1	49.6e	46.7
2001-10	48.9	45.6	49.1	50.8e	47.3

	Troy	Carmel/ Yorktown Heights	Peekskill	West Point	Little Falls	Indian Lake
1901-10		48.0		47.1	43.1	40.0
1911-20		47.9		47.6		
1921-30		48.5		48.1	44.9	41.0
1931-40		49.5		48.7	44.6	40.9
1941-50		49.2		48.5	44.6	39.5
1951-60		49.7	50.5	50.0	45.4	41.0
1961-70	48.1	48.7	48.7	49.0	44.0	39.6
1971-80	48.7	49.5	50.1	49.3	43.9e	39.7
1981-90	49.3	49.9	51.2	49.7	44.9	41.1
1991-00	49.6	50.4	50.1	49.4	44.3e	41.0
2001-10	49.8	51.4	52.4	52.6	45.7e	41.8

e = estimated

Ten year averages for eleven stations (Table 6) show that all stations except Glens Falls have their warmest decade in the 2001- -2010 period.

FIGURE 2:
Smoothed Annual Temperatures
Carmel / Yorktown, West Point, Mohonk Lake, Albany, Indian Lake

Figure 2 shows the average annual temperatures for the oldest Hudson valley U.S. Weather Service Cooperative stations with observation records of 100 or more years. The five station annual temperatures since 1900 have been averaged and smoothed by a five year running mean. The graph clearly shows a significant temperature rise of 3 degrees since 1980. The delayed cooling effect of eruption of Pinatubo in the early 1990s can also be seen.

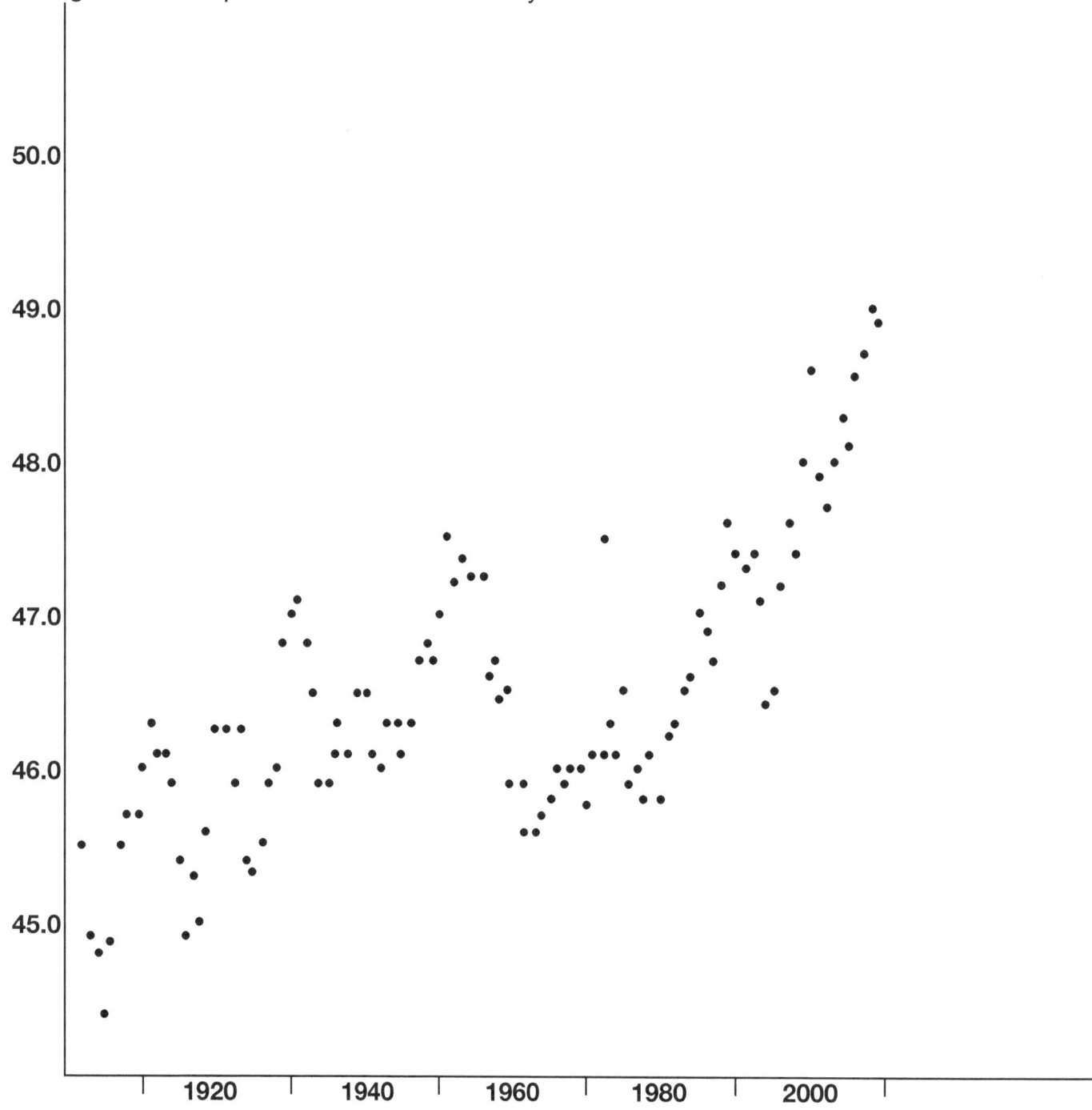

THE WARMEST YEARS

Divisional averages show that six or more of the ten warmest years have occurred since 2000 (Table 7). An examination of the warmest years (Table 8) for eighteen stations with 55 or more years of daily records shows that in the 31 year period since 1981, Little Falls has had seven of its warmest twelve years in that period. Three stations Poughkeepsie, Utica and Gloversville have seven of their warmest in that period; Albany and Newcomb have had eight of their warmest years. Four stations ; Millbrook, Glens Falls, Yorktown, and West Point, have had nine of their warmest years; Indian Lake and Alcove Dam have had ten of their warmest years; Troy has had eleven and two stations (Mohonk Lake & Saratoga Springs) have had their twelve warmest years in that 31 year period. All stations within that period of time show that the number of their warmest years is certainly greater than what would be expected by chance. At all weather stations, more than half of their warmest years have occurred in the ten year 2001-2010 period making that period the warmest decade since Hudson Valley temperature measurements began. The most recent year (2012) shows it to be the warmest at nine of fourteen stations and the second warmest at three.

TABLE 7
DIVISIONAL WARMEST YEARS

Hudson Valley		Mohawk Valley		Northern Plateau	
1998	51.0	1953	49.0	1998	44.5
1949	51.0	1998	48.7	2006	44.2
2006	50.7	1999	48.2	1949	44.1
1990	50.5	1949	48.0	2010	43.9
1991	50.4	1972	47.8	2011	43.6
1931	50.4	2006	47.4	2001	42.9
2010	50.0	1952	47.3	2005	42.8
1921	50.0	1938	47.1	1999	42.8
2011	49.2	2001	47.0	2002	42.3
2012	51.0	1991	47.0	2008	42.0
		1990	47.0	2012	44.7
		2012	47.8		

TABLE 8
WARMEST YEARS

	1	2	3	4	5	6	7	8	9	10	11	
Albany 1890-	51.7	50.5	50.2	49.9	49.6	49.4	49.1	49.0	48.9	48.8	48.4	
	2012	2006	2010	2011	2002	1999,01	1908	1998	1953	1949,80	1931	
Alcove Dam 1949 -	49.8	49.5	48.8	48.7	48.2e	47.9	47.8	47.6	47.5	47.3	47.0	
	2012	2010	2006	1998	1990	2002	1991,11	1999	2001	1953	1955	
Glens Falls 1893 -	49.4	49.0	48.9	48.7	48.4	48.1	47.7	47.6	47.1	46.8	46.5	
	1949	1953	1998	2012	2010	2006	1990	1999	2011	2005	2002	
Middletown 1951--2009	53.7	51.2	51.0	50.9	50.8	50.5	50.1	49.5	49.4	49.1		
	2006	1998	2002,10	1991	2007	1990	1999	1983	2001	1985		
Millbrook 1942-	51.7	51.4	51.1	50.9	50.7	50.6	50.5	50.4	50.3	49.4	49.2	
	2012	1998	2006	2002	1991,99	1990	2011	1988	2010	1973	1949	
Mohonk Lake 1891-	51.3	51.0	50.2	50.1	50.0	49.1	48.8	48.7	48.6	47.9	47.8	
	1998,12	2006	1999,01	2002	2010,11	2005	1990	1991	1949	2000	1995	
Peekskill 1946 -	54.6	54.0	53.9	53.8	53.4	53.0	52.8	52.7	52.6	52.5	52.3	52.0
	2012	2010	1949	2006	2011	1953	1991	1990	2001	2002	2008	2005
Poughkeepsie 1928 –	53.8	53.7	53.2	53.1	53.0	52.1	51.8	51.6	51.2	50.8	49.9	
	2006	2012	1953	1949	1990	2005,11	2010	1998	1941	1995	1931	
Saratoga Spgs 1941-	49.4	48.6	48.2	47.7	47.1	47.0	46.8	46.7	46.6	46.4	46.1	45.8
	2010	2011	2012	1998	1990	1991	2001,06	2002	1999	2005	1949	2000
Troy Lock & Dam 1956-	52.9	52.7	51.7	51.6	51.5	51.3	51.2	51.1	50.8	50.4	50.3	50.0
	2012	1998	2006	1990	1999	2001	2002	1991	2011	2005	2010	1995
West Point 1890 -	53.8	53.4	53.3	53.1	52.6	52.4	52.3	50.8	50.7	50.6	50.5	50.4
	2010,12	2006	2002	2011	2001	1991	2005	1931	1949	1921	1938	1990
Westchester Airport 1949-	54.7	54.4	54.0	53.8	53.2	53.1	53.0	53.0	52.9	52.6	52.3	
	1998	2012	1999	2006	2010	1953	1990	1991	1983,11	1985	2002	
Yorktown 1888-	54.6	53.8	53.2	52.6	52.2	52.0	51.9	51.8	51.7	51.6	51.5	
	2012	2006	1998	2010,11	1991	1999	1900	1990	1949	2001	2005	

	1	2	3	4	5	6	7	8	9	10		
Gloversville 1892 - 2007	47.9 1998	47.7 1953	47.4 1949,06	47.2 1921	47.0 1991	46.9 1999	46.8 2005	46.7 1990	46.6 1930,73	46.5 1938,02		
Little Falls City Res. 1897 –	49.3 2005	48.2 1998	48.0e 1999	47.6 2011	47.3 1990,10	47.2 1953,06	47.1 1949	46.9 1921	46.8 1938	46.7 1915		
Utica 1927- 06	50.9 2006	50.7 1998	49.7 1973	49.6 1999	48.9 1953	48.3 1991	48.1 1949,84	48.0 1990	47.9 2002	46.3 1947		
Indian Lake 1899-	44.0 2012	43.8 1998	43.7 1990	43.2 2011	43.1 1921,53	43.0 2006	42.9 2010	42.8 1949	42.6 1991	42.4 2002	42.3 2001	42.2 1999
Newcomb 1940-	44.4 2011	44.0 1953	43.9 2006	43.8 1957	43.7 1998	43.1 2010	43.0 1949	42.8 1955	42.6 2005	42.2 1990	41.9 2001	41.6 1991

WARMEST SEASONS

Winter

Of their ten warmest winters of record, (Table 9) three stations (Mohonk Lake, Poughkeepsie, Yorktown) have had five in the 1981 – 2010 period; four stations (Albany, Peekskill, West Point, Indian Lake) have six; and four stations (Saratoga Springs, Troy, Little Falls, Utica) have seven. All had a greater number than would be expected by chance in that 30 year period. The recent 2011 - 2012 winter has been the second warmest for nearly all stations.

Summer

Of their hottest summers of record, (Table 10) one station (Indian Lake) has four of its warmest ten summers in the 1981-2011 period of record; two stations (Glens Falls, Little Falls) have five; three stations (Albany, West Point, Utica) have seven, two stations (Saratoga Springs & Troy) have eight; one station (Yorktown) has eleven, two stations Mohonk Lake has ten; Poughkeepsie and Peekskill have eleven. As can be seen nearly all of these warmest summers have come in the 2001-2011 period.

Spring

Of the ten warmest springs of their record, two weather stations (Poughkeepsie & Little Falls) have four of their warmest in the 1981-2011 period, one station (Glens Falls) has five warmest, two stations (West Point & Yorktown) have six warmest, three stations (Millbrook, Saratoga Springs & Indian Lake) have seven, three stations (Albany, Mohonk Lake & Troy) have eight, two stations (Middletown & Peekskill) have eight and nine of their warmest, and one station (Alcove Dam) has nine of its warmest ten springs in that period. About half of their warmest springs for all stations have occurred since 2000.

Autumn

Of the ten warmest autumns of record, two stations Glens Falls & West Point have two of their warmest ten in the 1981-2011 period; five stations (Albany, Millbrook, Poughkeepsie, Little Falls, Indian Lake) have four of their warmest ten; four stations (Alcove Dam, Middletown, Mohonk Lake, Yorktown) have five of their warmest; two stations (Saratoga Springs & Westchester Airport) have six of their warmest; one station (Peekskill) has seven and another station (Troy) has eight. About half of these warmest cited have occurred since 2000.

In summary it can be seen from the temperature record for all time periods, the 30 year span ending in 2010 has been the warmest for nearly all Hudson Valley climate divisions and long term individual stations. The warming has continued beyond the end of 2010 as the year 2011 ranks within the warmest ten for all stations and 2012 ranks as the warmest year since records began at all stations. The winter of 2011 – 2012 has also been in the warmest ten. The month of March 2012 for nearly all stations has been the warmest of record. April 2012 can also be included as one of the warmest ten of record. The spring of 2012 is also one of the warmest ten at nearly all stations. From January 2011 through October 2012 average monthly temperatures have been above normal (1981 – 2010) average.

TABLE 9
WARMEST SEASONS - WINTER
December (of previous year), January, February

Albany City /Airport 1890 –

1	2	3	4	5	6	7	8	9	10	
32.3	31.6	30.2	29.6	29.4	28.9	28.7	28.6	28.5	28.2	28.1
2002	2012	1998	1932	1949	1991,97	2006	1995	1937	1983	1890

Alcove Dam 1949-

31.4	30.9	29.6	28.8	28.3	27.9	27.3	26.9	26.6	25.3	23.6
2002	2012	1949	1998	1991	1953	2006	2001	1999	1984	1990

Glens Falls Farm / Airport 1948 -

29.8	29.4	29.0	28.7	27.1	26.4	26.0	24.5	24.4	22.4
1953	1949	2002	1998,12	1950	1999	1991	1990	2006	1984

Indian Lake 1900 –

25.8	23.7	23.5	23.4	22.5	22.4	22.3	22.0	21.8	21.2
2002	1998,12	1949	1933	1932	1991	1953,99	1937	1913,83	2006

West Point 1890 –

36.0	35.8	35.4	33.9	33.6	33.5	33.2	33.1	33.0	32.8	32.4
2012	1932	2002	1937	1890	1953,91	1998	1933	2006	1983	1999

Mohonk Lake 1896 –

33.6	32.4	32.3	30.9	30.3	29.8	29.7	29.6	29.1	29.0	28.7
2002	1932	2012	1998	1949	1933,99	1937,53	1991	1913,19	2006	1997

Peekskill 1946 -

37.2	36.8	35.2	35.0	34.3	33.8	33.5	33.4	33.1	32.8	32.7
2002	2012	1953	1949	1975	1991	1998	1954	2006	2007	2000

Poughkeepsie 1889 –

35.7	34.1	33.5	33.1	33.0	32.8	32.7	32.2	31.9	31.6
2012	2002	1932	1937	2006	1997	1998	1953	1890	1949

Saratoga Springs 1941 –

30.0	28.1	27.4	26.9	26.4	26.3	26.2	26.1	23.2	22.5
2002,12	1949	1998	1919	1954	2006	1991	1997	1975	1990

Carmel / Yorktown Heights 1889 –

1	2	3	4	5	6	7	8	9	10	
36.7	36.4	36.3	35.7	35.0	34.8	34.0	33.9	33.5	33.3	32.9
2012	2002	1890	1932	1937	1998	1953	1933,06	1949	1991,2013	1999

Troy 1896 –

33.9	33.4	31.3	30.5	30.4	30.1	30.0	29.9	29.8	29.5
2002	2012	1998	2006	1991	1953	1997	1975	1999	1932

Gloversville 1893 – 2006

29.6	27.8	27.7	27.4	27.3	26.9	26.2	26.1	26.0	25.6
2002	1933	1937,53	1998	1932	1949	1983	1919,99	1975	1991

Little Falls City Reservoir 1898 –

30.4	28.8	26.9	26.8	26.6	26.2	25.6	25.4	25.2	25.1	24.8
2002	2012	1933	1983	1932	1998	1953	1991,99	1997	1949	2005,06

Utica 1893 - 2006

31.1	29.5	29.1	28.7	27.5	27.4	27.1	27.0	26.7	26.5
2002	1932	1998	1933,37	1997	1999	1949	1991	1983	2006

TABLE 10
SEASONS
WARMEST SUMMERS

	1	2	3	4	5	6	7	8	9	10	
Albany C/ AP 1890 -	73.5 2005	73.0 1949	71.5 2010,12	71.3 1900,55	71.2 2011	71.1 1937	71.0 2002	70.9 1938,73	70.8 2006,08	70.7 1999	70.4 2003
Alcove Dam 1949-	72.7 1949	70.3 2005	69.7 1955	69.4 1999	69.1 2010	68.4 2012	68.3 1984	68.0 2006	67.6 2011	67.3 1998	67.1 2001
Mohonk Lake 1891-	72.8 2005	72.0 2002	71.9 1999	71.8 2010	71.1 1988	71.0 2001	70.8 2011	70.7 2012	70.5 2006	70.4 1949	70.3 1995
Peekskill 1946-	74.8 2010	74.6 1949	74.5 2005	74.4 1999	73.2 1988,11	73.1 1983	73.0 2012	72.8 1981	72.7 1955,80	72.5 1984,02	72.1 1987 72.0 2001,07
Glens Falls C&AP 1893	72.6 1949	72.5 1955	71.3 2005	70.8 1944	70.2 1973	70.1 1943	70.0 1959	69.7 2011	69.0 1999	68.8 2010	68.3 2012 67.8 2002
Middletown 1951- 2010	73.3 2005	73.2 1999	72.5 1983	72.3 2010	72.2 2002	72.1 2006	71.6 1988	70.5 1987	70.3 1981	70.2 2001	
Poughkeepsie 1889 -	74.2 2005	72.7 2006	72.2 2010	72.1 2008,12	72.0 2011	71.7 2007	71.3 1991	71.2 1949,95	70.9 1973	70.7 1994	70.3 2004
Saratoga Spgs 1903-	72.4 2005	71.7 1949	71.1 2010	70.9 2012	70.6 1955	70.2 1933	70.1 1973,11	69.9 1999	69.8 1959,95	69.5 1988	69.4 2002,03 69.3 2007
Troy 1895 -	74.7 1933	74.0 2005	73.2 1973	72.8 1949	72.6 1999,12	72.5 1995	72.4 2002	72.2 2010	72.0 1937	71.9 2006	71.8 1983,91 71.7 2011
West Point 1890-	75.2 2005	75.0 1966	74.9 1955	74.7 2010	74.4 1973	74.2 1949	74.1 2002	74.0 1999	73.9 1959,83	73.8 1943	73.7 1991 73.6 2011,12

	1	2	3	4	5	6	7	8	9	10			
Yorktown Hts 1888-	73.3	73.0	72.8	72.2	71.0	71.8	71.7	71.6	71.5	71.4	71.3	71.2	
	2005	1999	2010	1949	2006,08	1988,11	1998	1939,55	2012	1991	2001	2002	
Gloversville 1892-2006	72.0	70.0	69.4	68.8	68.5	68.3	68.2	68.1	68.0	67.9			
	2005	1949	1937	1973	1938	1959	1901,44	1943	1988,95	1983,99			
Little Falls Cs. 1897-	71.4	70.1	69.4	69.1	68.8	68.7	68.6	68.5	68.4	68.3	68.2	68.0	
	2005	1949	1938,73	1955	2011	1939,95	1901	1898	2002	1936	1999	2012	
Utica 1891-	72.3	71.6	70.6	69.8	69.7	69.6	69.5	69.4	69.1	68.7			
	2006	1949	2005	1955	1999	1995	2002	1959,73	1937	1991	1987,88,94		
Indian Lake 1900 -	65.8	65.1	64.8	64.5	64.3	64.2	64.1	63.8	63.7	63.4	63.3		
	2005	1955	1901	1949	2011	1900	1921	1906	1947,88	1959	1999		

TABLE 13
WEST POINT
1821-2012
Warmest Seasons
Degrees F

	1	2	3	4	5	6	7	8	9	10	
WINTER (D,J,F)	36.9	36.1	35.3	33.6	33.5	33.4	33.3	33.2	33.0	32.9	
	2002	2012	1932	1913	1991	1924	1998	1890,37	2006	1983	
SPRING	54.7	54.6	54.1	53.5	52.4	52.3	52.1	52.0	51.8	51.6	51.5
	2010	2012	1921	1991	1942,77	1985	1925	1955	1918	2004	1986
SUMMER	75.2	75.0	74.9	74.7	74.4	74.2	74.1	73.9	73.8	73.7	73.6
	2005	1966	1955	2010	1973	1949	1999,02	1959,83	2012	1943,91	2011
FALL	57.7	57.5	56.8	56.7	56.6	56.2	56.1	55.9	55.8	55.7	55.6
	1921	2005	2007	1946	1961	1941	1963	1968	1912	1971	1973

TABLE 14
PEEKSKILL
1946 - 2012
Warmest Seasons

	1	2	3	4	5	6	7	8	9	10	
SUMMER	74.8	74.6	74.5	74.4	73.2	73.1	72.8	72.7	72.5	72.1	72.0
	2010	1949	2005	1999	1988,11	1983	1981	1955,80	1984,02	1987	2001
WINTER D, J, F (Dec. of previous year)	37.2	36.7	35.2	35.0	34.3	33.8	33.5	33.4	33.1	32.8	32.7
	2002	2012	1953	1949	1975	1991	1998	1954	2006	2007	2000
SPRING M,A.M	55.5	55.4	51.8	52.9	52.5	51.7	51.6	51.4	51.2	51.1	51.0	50.7
	2010	2012	1986	1991	1985	2000	1977	2009	2006	2004	1946	1998,08
FALL S, O, N	57.5	57.1	56.7	56.3	55.8	55.5	54.8	54.7	54.4	54.0	
	1946	1948	2007	1999	2001	1985	1983	1971	1990	2010	

TABLE 15
MIDDLETOWN
1951 - 2012
Warmest Seasons
(December of previous year)

	1	2	3	4	5	6	7	8	9	10
D, J, F	37.2	33.0	32.9	32.5	30.8	30.7	30.5	30.3	29.9	29.4
	2006	1990	1998	2002	1991	1953	1999	1954	2000	1997
M,A,M	53.0	51.2	50.7	50.6	50.0	49.8	49.2	47.5	47.2	46.5
	2012	1977	1998	2010	1991	1985	2004	1955	2004	2002
J, J, A	73.3	73.2	72.5	72.3	72.2	72.1	71.6	70.5	70.3	70.2
	2005	1999	1983	2010	2002	2006	1988	1987	1981	2001
S, O, N	56.2	56.1	55.9	54.0	53.9	53.8	53.7	53.6	53.4	53.2
	2005	2007	1961	75,2001	1971	1990	1963	2006	1970	1999

TABLE 16
MILLBROOK
1942 -2012
Warmest Seasons

Winter

1	2	3	4	5	6	7	8	9	10	
34.1	34.0	32.9	32.7	31.8	31.0	30.9	29.8	28.2	27.0	
2002	2012	1998	1953	1997	1949	1999	2006	1954	1990	

Summer

72.3	71.5	71.2	70.7	70.4	70.2	70.1	69.5	69.1	68.4	
2005	2008	2002	2010	2006,11	1949	2012	2007	2004	2001	

Spring

54.0	53.0	52.6	52.1	51.1	50.5	49.8	49.2	48.0	46.6	44.9
1998	2012	1945	2010	2004	1946	2001	1991	1977	2007	1963

Fall

55.7	54.7	54.6	54.0	53.7	53.3	52.5	52.2	51.6	49.8	
2005	1999	2011	2001	1961	2007	2010	2006	1948	1949	

TABLE 17
MOHONK LAKE
1891 - 2012
Warmest Seasons/Years
Degrees F

	1	2	3	4	5	6	7	8	9	10	11	12
WINTER (Dec. Jan. Feb)	33.4 2002	32.3 2012	32.0 1932	30.7 1998	30.0 1949	29.6 1999	29.5 1937,91	29.3 1953	29.0 2006	28.8 1919,07		
SPRING M,A,M	52.6 2012	52.3 2010	49.8 1921,91	49.2 1998	48.8 1977,04	48.5 1985	48.3 2000	48.2 1986	48.1 1903	47.9 1910	47.8 1955	47.7 2006
SUMMER J,J,A	72.8 2005	72.0 2002	71.8 2010	71.1 1988	71.0 2001	70.8 2011	70.7 2012	70.5 2006	70.4 1949	70.2 1995	70.1 1991	70.0 2008
FALL S,O,N	54.3 2001	53.7 2005	53.5 1948	53.0 1999,07	52.7 1961	52.6 1941	52.5 1927,63	52.4 1990	52.3 1900	52.2 1998	51.3 2010	
ANNUAL	51.3 1998,12	51.0 2006	50.2 1999,01	50.1 2002	50.0 2011	49.1 2005	48.8 1990	48.7 1991	48.6 1949	47.9 2000	47.8 1995	

TABLE 18
POUGHKEEPSIE
Warmest Seasons/Years

	1	2	3	4	5	6	7	8	9	10	11
D, J, F	35.6 2012	34.1 2002	33.5 1932	33.1 1937	33.0 2006	32.8 1997	32.7 1998	32.2 1953	31.9 1890	31.6 1949	
M,A,M	54.3 2012	52.7 2010	52.6 1941	52.4 1991	52.3 2004	52.2 1977	51.8 1945	50.6 1955	49.6 1946	48.7 1944	47.9 1965
J, J, A	74.2 2005	72.7 2006	72.2 2010,11	72.1 2008,12	71.7 2007	71.3 1991	71.2 1949,95	70.9 1973	70.7 1994	70.3 2004	
S, O, N	57.7 2007	57.6 1961	57.4 2005	56.6 1931	56.3 1963	54.7 1948	53.6 1973	53.1 1999	53.0 1949	52.8 2001	
ANNUAL	53.8 2006	53.7 2012	53.2 1953	53.1 1949	53.0 1990	52.1 2005,11	51.8 2010	51.6 1998	51.2 1941	50.8 1995	

TABLE 19
ALBANY CITY / AIRPORT
1890 – 2012
Warmest Seasons/Years
Degrees F

	1	2	3	4	5	6	7	8	9	10	
WINTER (December of previous year)	32.3 2002	31.6 2012	30.2 1998	29.4 1953	29.1 1949	28.9 1991,97	28.7 2006	28.6 1995	28.2 1983	27.9 1999	27.8 1954
SUMMER	73.5 2005	73.0 1949	72.4 1955	71.8 1959	71.5 2010,12	71.4 1973	71.3 1943	71.2 1938,11	71.0 1944,02	70.9 1939	70.8 2006,08
SPRING M,A,M	52.4 2012	51.6 2010	50.6 1991	50.0 1998	49.7 1986	49.6 1942,45	49.5 2004	49.3 1987	49.0 1985	48.7 1973	48.2 2000
FALL	54.1 1961	53.6 1946	53.5 2005	53.0 2001	52.9 2007	52.7 1970,75	52.5 1999	52.3 1941,48	52.2 1963,90	52.1 1971	
ANNUAL 1890-	51.7 2012	50.46 2006	49.9 2011	49.6 2002	49.4 1999,01	49.1 1908	49.0 1998	48.89 1953	48.81 1949,80	48.41 1931	48.35 1991

TABLE 20
ALCOVE DAM
Warmest Seasons

	1	2	3	4	5	6	7	8	9	10
D, J, F	31.4 2002	31.0 2012	29.6 1949	28.8 1998	28.3 1991	27.9 1953	27.3 2006	26.9 2001	26.6 1999	25.3 1984	23.6 1990
M,A,M	50.4 2010	49.9 2012	47.8 1991	47.3 1998	47.1 1977	47.0 1987	46.6 2004	46.4 1985	46.2 2009	45.9 2000	44.9 2002
J, J, A	72.7 1949	70.3 2005	69.7 1955	69.4 1999	69.1 2010	68.3 1984,12	68.0 2006	67.6 2011	67.3 1998	67.1 2001	67.0 1987
S, O, N	53.4 1999	53.2 1961	51.8 2005	51.5 1948	51.3 2007	51.2 1959	51.1 2001	50.8 1963	50.2 2009	49.7 1971	

TABLE 21
TROY
Warmest Seasons/Years

	1	2	3	4	5	6	7	8	9	10	
D, J, F	33.9	33.5	31.3	30.5	30.4	30.1	30.0	29.9	29.8	29.5	
	2002	2012	1998	2006	1991	1953	1997	1975	1999	1932	
M,A,M	52.6	52.2	51.4	50.8	50.3	49.8	49.7	49.2	49.1	47.8	47.0
	2012	2010	1998	1991	1986	2000	1921	2004	1977	1968	2008
J, J, A	74.7	73.9	73.1	73.0	72.8	72.5	72.3	72.2	71.9	71.7	
	1933	2005	1973	1999	1995,12	2002	2010	1988	1991	2011	
S, O, N	55.9	55.2	55.1	54.4	54.3	54.1	53.8	53.7	53.6	52.6	
	2001	1999	2005	2007	1991	1961	1971	2009	1990	2002	
ANNUAL	52.9	52.7	51.7	51.6	51.5	51.3	51.2	51.1	50.8	50.4	50.3
	2012	1998	2006	1990	1999	2001	2002	1991	2011	2005	2010

TABLE 22
GLENS FALLS (Fire Station & Farm)
Airport 1948 - 2012
Warmest Seasons/Years

	1	2	3	4	5	6	7	8	9	10	
D, J, F	29.8	29.4	29.0	28.7	27.1	26.4	26.0	24.5	24.4	22.4	
	1953	1949	2002	1998,12	1950	1999	1991	1990	2006	1984	
M,A,M	50.0	49.2	49.1	48.8	48.7	48.6	48.5	48.4	47.3	45.0	
	1946	2012	2010	1921	1998	1991	1945	1903	1986	1911	
J, J, A	72.6	72.5	71.3	70.8	70.2	70.1	70.0	69.7	69.0	68.3	
	1949	1955	2005	1944	1973	1943	1959	2011	1999	2010	
S, O, N	54.0	52.8	52.7	52.6	52.4	51.9	51.5	51.4	51.2	50.7	
	1946	1961	1948	1900	1947	1949	1999	1954	1963	2005	
ANNUAL	49.4	49.0	48.9	48.7	48.4	48.1	47.7	47.6	47.1	46.8	46.5
	1949	1953	1998	2012	2010	2006	1990	1999	2011	2005	2002

TABLE 23
SARATOGA SPRINGS
Warmest Seasons/Years

	1	2	3	4	5	6	7	8	9	10		
D, J, F	30.1	30.0	28.1	27.4	26.4	26.3	26.2	26.1	25.9	23.2	22.5	
	2012	2002	1949	1998	1954	2006	1991	1997	1999	1975	1990	
M,A,M	51.8	51.3	49.0	48.3	48.2	48.0	47.3	47.1	47.0	45.9	45.6	
	2012	2010	1991	2004	1977	1998	2002	1945	2000	1942	1946	
J, J, A	72.4	71.7	70.8	70.6	70.2	70.1	69.7	69.6	69.5	69.4	69.3	
	2005	1949	2010,12	1955	1933	1973,11	1959	1999	1988,95	2002,03	2007	
S, O, N	52.6	51.3	51.1	50.8	50.6	50.5	50.4	49.8	49.7	49.1		
	1971	1961	2005	2009	2001	1999	1948	2007	1947	2002		
ANNUAL	49.4	48.6	48.2	47.7	47.1	47.0	46.8	46.7	46.6	46.4	46.1	45.8
	2010	2011	2012	1998	1990	1991	2001,06	2002	1999	2005	1949	2000

TABLE 24
LITTLE FALLS - CITY RESERVOIR
Warmest Seasons

	1	2	3	4	5	6	7	8	9	10	
D, J, F	30.4	28.6	26.9	26.8	26.6	26.2	25.6	25.4	25.2	25.1	24.8
	2002	2012	1933	1983	1932	1998	1953	1991,99	1997	1949	2005,06
M, A, M	49.6	48.4	48.1	48.0	47.8	47.3	47.0	45.2	44.1	43.3	
	2012	2010	1998	1921	1903	1945	1942	1946	1941	2005	
J, J, A	70.7	70.6	70.4	69.3	69.2	69.0	68.5	68.4	68.3	68.1	
	1949	2005	1999	1955	1937	2011	1973	1943	1921	2010	
S, O, N	52.8	52.3	51.7	51.6	51.4	51.3	51.2	51.1	50.3	50.2	
	1999	2007	1946	1913	2001	1971	2005	1931	1948	1947	

TABLE 25
INDIAN LAKE
1899 - 2012
Warmest Seasons/Years

	1	2	3	4	5	6	7	8	9	10	
WINTER	25.8	23.7	23.5	23.4	22.5	22.4	22.3	22.1	22.0	21.8	21.2
(Dec, Jan	2002	1998,11	949	1933	1932	1991	1953	1999	1937	1913,83	2006
Feb)											
SPRING	44.4	44.3	44.2	42.8	42.6	41.6	41.3	41.1	40.9	40.8	
	2012	1921	2010	1986	1991	1977,98	1903	1910	2004	1987	
SUMMER	65.8	65.1	64.8	64.5	64.2	64.1	63.8	63.7	63.4	63.3	
	2005	1955	1901	1949,11	1900	1921	1906	1947,88	1959	1999,12	
FALL	47.7	47.5	47.3	47.2	47.0	46.7	46.6	46.4	46.2	46.0	
	1961	1931	2005	1953	2001	1927,71	2007	1990	1946	1913,48	
ANNUAL	44.0	43.8	43.7	43.2	43.1	43.0	42.9	42.8	42.6	42.4	42.3
	2012	1998	1990	2011	1921,53	2006	2010	1949	1991	2002	2001

WARMEST MONTHS

The warmest months of the periods of record for all Hudson Valley watershed stations show greater heat than what would be expected statistically for the 1981-2010 period. (See Tables 26 - 40).

For Albany City/ Airport and West Point dating back to 1890, ten months of the year show at least two months in the warmest ten. For Alcove Dam dating back to 1949, all months of the year had at least four in the 30 year 1981-2010 span. At Glens Falls with its 106 year combined record there were ten months with greater than three in the 1981-2010 period. At Middletown there were at four to eight months falling in the ten warmest of record in the 1981-2010 period. At Millbrook all months except October had five to eight warmest in at period. At Mohonk Lake with 121 years of record, ten months had the record warmest greater than three in the 30 year period. At Peekskill only October had as few as three of its hottest individual months falling in the 1981-2010 period. Poughkeepsie with its combined 110 year record had two to six months with the warmest falling in the 1981-2010 period. At Saratoga Springs and Troy, for all months at least half the ten warmest fell in the 1981-2010 period. At the combined Carmel/Yorktown Heights record dating back to 1888, only March has as few as 3 months within the hottest ten of record in the 1981-2010 period. December had as many as nine. Little Falls and Indian Lake with their 112 year record only June came with two months of their warmest ten in the 1981 -2010 period. March 2012 was the warmest third month at all but two stations.

Of even greater significance is the period of 2001 through 2012 when thirteen stations had at least one of their warmest in each month. The average for those thirteen was three.

WARMEST DAYS

An examination of maximum daily temperatures for each month (Table 41) at sixteen stations shows that in the 1981 – 2010 period January had nine; February had eleven; March had eleven; April had eight; May had two; June had three; July had two; August had three; September had two; October had none; November had five; December had thirteen.

It appears that that there were a greater frequency of record breaking monthly daily maxima in the six coldest months of the year in the 1981 – 2010 period. The cause of this difference is uncertain.

TABLE 26
WARMEST MONTHS / YEARS
Westchester Airport
1949 - 2012

	1	2	3	4	5	6	7	8	9	10	11
January	37.5	37.1	36.5	36.2	35.0	34.8	34.5	34.4	33.8	33.0	32.9
	1998	1990	2006	2002	1995	1953	2007	2012	1967	1989	2008
February	38.5	38.0	37.9	37.6	36.8	36.2	35.9	35.2	35.1	35.0	34.4
	1998	1997	1984	2012	1954	2002	1991	1981	1953,90	1976	2005
March	47.8	44.8	44.5	42.7	42.6	42.5	42.4	42.3	41.5	41.4	
	2012	2010	2000	1985	1973	1983	1998	1995	1991,06	1977	
April	54.0	53.1	52.4	52.2	51.8	51.6	51.5	51.4	51.3	51.2	51.0
	2010	1985	1991	1968	1952	1974,08	2012	1969	2002	2006	1994
May	65.1	64.2	62.9	62.8	62.6	62.5	62.4	62.2	62.0	61.9	61.7
	1991	1965	2012	1959	1964	1985	1998	2010	1955	1982	1986,04
June	71.3	71.0	70.8	70.7	70.4	70.2	70.0	69.9	69.7	69.6	69.5
	1999	2005	2008	2010	1994	1967	1983	1976	1984,96	1966	2001
July	77.2	77.1	77.0	76.9	76.8	76.4	76.2	76.0	75.7	75.5	75.3
	1999	1955	2010	2011	1966	1983	1952,12	1994,95	1968	1993	2006
August	75.6	75.5	75.3	75.2	75.1	74.9	74.7	74.1	74.0	73.9	73.0
	1983	1993	2005	1955	1970	1959	2001	1998	2003	1988	2012
September	71.1	68.7	68.4	68.2	68.0	67.9	67.7	67.6	67.5	67.2	
	1961	1959	2005	1983	1971	1998	2010,11	1985	1968	1999	
October	60.3	60.1	58.6	58.0	57.9	57.8	57.7	57.1	56.3	56.2	
	1971	2007	1963	1954	1995	1984	1990	1968	1955	1970	
November	49.2	48.8	48.5	48.0	47.5	47.4	46.8	46.3	46.2	45.9	45.8
	1994	2006	2011	1999	1963	2004	2003	1964	1966,05	1985	1982,90
December	41.4	40.4	40.1	39.7	38.5	38.3	38.2	37.6	37.5	37.0	36.8
	1998	2006	1984	2011	1994,96	1982	2012	1971	1999	1953	1956
Annual	54.69	54.43	53.99	53.75	53.12	53.06	53.03	53.01	52.88	52.87	52.58	52.29
	1998	2012	1999	2006	2010	1953	1990	1991	2011	1983	1985	2002

TABLE 27
WARMEST MONTHS / YEARS
CARMEL 1888 - 1980; YORKTOWN HEIGHTS 1965 - 2012

	1	2	3	4	5	6	7	8	9	10	11
January	38.6 1932	37.0 2006	35.7 1937	35.6 1890	35.3 1933	35.2 1950,90	35.0 1913	34.5 2007	34.4 1998,02	33.4 2012	
February	37.4 2012	36.4 1998	36.3 1890	35.7 2002	35.1 1981,84	35.0 1991	34.9 1954	34.8 1953	33.8 1997	33.6 1990	33.2 2006
March	48.4 2012	45.5 1946	45.4 1903	44.4 2010	43.9 1921	43.7 1945	43.3 1898	43.0 1977,00	42.5 1936	42.4 1973	
April	54.3 2010	53.4 2012	53.3 1941	52.7 2006	52.6 1921	52.4 1945	52.1 2008	51.9 1991	51.8 1979	51.4 1968	51.2 2009	51.0 1994
May	64.4 1896	64.2 1991	63.5 2012	63.1 1911	62.8 1944	62.6 2010	62.5 2007	62.2 1998,04	62.1 1975	62.0 1904	61.9 1903
June	71.8 1925	71.5 1892	71.2 1895	71.0 1899	70.9 1943	70.8 2005	70.7 2008	70.2 1894	70.0 1999	69.8 2001	69.5 2010
July	77.5 1999	76.1 2010	76.0 1955	75.8 2011,12	75.2 1901	75.0 2006,08	74.4 1988	74.3 1994	74.2 2005	73.9 2002	
August	74.8 2005	74.6 2001	74.4 1955	73.9 1939	73.8 2012	73.4 1988	73.2 1937	73.0 1980,98,03	72.9 2010	72.5 1959	72.3 1973	72.2 1984,07
September	69.4 2005	68.5 2007	68.1 1961	67.7 1921	67.5 1900	66.9 2010	66.7 1891	66.4 1915	66.1 1894	66.0 2002	65.9 1971
October	61.3 2007	58.7 1971	58.2 1947	57.8 1900,20	57.6 1954	57.4 1949	57.0 1995	56.8 1990	56.3 2012	56.2 1963	55.7 1984
November	49.9 2006	48.9 2009	48.7 2011	47.8 1975	47.5 2001	46.8 1979	46.5 1994	46.4 1999	46.3 2005	45.6 1946	45.5 2003
December	40.9 2006	39.3 1891	39.1 2001	38.7 2012	38.3 1998,11	37.4 1984	37.0 1990	36.5 2007	36.4 1996	36.3 1993	36.0 1982
Annual	54.62 2012	53.78 2006	53.19 1998	52.63 2011	52.57 2010	52.54 2007	52.21 1991	51.98 1999	51.86 1900	51.84 1990	51.71 1949	51.67 2001

TABLE 28
WARMEST MONTHS / YEARS
PEEKSKILL
1946 - 2012

	1	2	3	4	5	6	7	8	9	10	11
January	36.7	36.2	35.8	35.2	34.9	34.7	34.4	33.7	33.6	33.4	
	1990	2006	2001	1950	1949	2002	1975	1953	2007	2012	
February	37.6	37.4	36.8	36.0	35.9	35.6	34.9	34.5	34.3	34.1	
	1959,12	2002	1954	1953	1981	1984,98	1949	1976	1997	2000	
March	48.8	46.9	45.8	44.8	43.2	42.9	42.5	42.0	41.4	41.3	41.2
	2012	1946	2010	2000	1977	1991	1973	1985	1990	1998	1995
April	56.2	53.7	52.9	52.6	52.5	52.2	52.1	52.0	51.9	51.8	51.6 51.4
	2010	2008	2012	2002,05,06	1985,11	1952	2009	1981	1999	1949,76	1991 1994,01
May	64.4	64.2	64.0	63.8	63.6	63.2	63.1	63.0	62.8	61.9	61.6
	2010,12	1991	1975	1972	1986	2011	2004	1985	1980	1955	1999,01
June	73.0	71.9	71.6	71.4	71.2	71.1	70.9	70.7	70.6	70.4	70.3
	2010	2005	2008	1984	1983,99	1949	1957	1989	2011	2007	2001
July	78.8	78.2	78.1	77.8	77.0	76.6	76.5	76.1	76.0	75.8	75.6
	1999	2010	1961	1949	1955	1981,88	2011,12	2006	1995	1980	1987,08
August	76.1	75.8	75.5	74.8	74.4	74.3	74.0	73.8	73.7	73.3	
	2002,05	1980	2001	1949,55	2010	1988	2003	2009,12	1984	1983	
September	68.0	67.7	67.6	67.5	67.4	67.2	67.1	67.0	66.9	66.6	66.5
	2010	2007	1983	2005	1999	1946	2011	1980	1948,08	1985	1971
October	60.9	60.2	59.5	58.9	58.4	57.5	57.4	56.8	56.6	56.5	
	2007	1949	1947	1954	1946	1984	1971	1963	1990	1955	
November	50.8	48.6	48.4	48.2	47.4	47.0	46.8	46.4	46.2	45.3	45.0
	1948	2011	2001	1999,06,09	1975	1950	1946	1994	1963	1985	2003
December	39.6	39.3	38.0	37.2	37.0	36.8	36.5	36.2	35.9	35.8	
	2001,06	2011	1984	1953	1998	1990,99	1982	1994	1974	1952	
Annual	54.68	54.0	53.86	53.78	53.42	52.98	52.77	52.69	52.67	52.42	52.36
	2012	1949	2006	2011	1953	1991	1990	2001	2002	2008	2007

TABLE 29
WEST POINT
1821 - 2012
Corrected for Time of Observation (1890 - 2012)
Warmest Months

	1	2	3	4	5	6	7	8	9	10		
January	36.7	35.8	34.9	34.6	34.4	34.2	33.8	33.6	33.5	33.4	33.1	
	1932	1990	1937	1880	2002	1913	1933	2006	1950	1858	2012	
February	37.2	36.7	36.1	36.0	35.3	35.1	35.0	34.8	34.5	34.0	32.8	
	2002	2012	1954	1828	1834	1981	1842	1821	1976	1991	1984	
March	48.1	45.4	45.2	45.0	44.8	44.4	44.0	44.2	43.9	43.5	43.0	
	2012	1946	1945	1921	2010	1898	1903	1825	2000	1973	1871	
April	55.8	54.9	54.2	53.9	53.6	53.1	52.8	52.7	52.6	52.4	52.1	
	2010	1921	1878	1968	1871	1941	2006	2002	1916	2008	2005	
May	66.0	65.5	65.0	64.8	64.6	64.4	64.2	63.7	63.5	63.3	63.0	
	1991	1965	1826	1880	1872	1864	2012	2004	2010	1896	1944	
June	75.0	74.7	74.0	73.7	73.5	73.2	73.0	72.9	72.3	72.2		
	1831	1876	1865	1869	1825	1872	2005	1828	1943	1838,83		
July	78.6	78.2	77.5	77.3	77.1	77.0	76.9	76.8	76.7	76.2	76.1	76.2
	1876	2010	1824	1872	2011	1868	1823	2012	1856,78	2006	2005	1955
August	76.7	76.2	75.7	75.6	75.5	75.4	75.2	75.1	75.0	74.9	74.3	
	2005	2001	2002	1876	1908	1821	2003	1877	1955,12	1830,64	2010	
September	70.7	70.0	69.4	68.6	68.1	67.9	67.8	67.6	67.5	67.2	67.0	
	2005	1961	1865	2010	2011	2001	1921	2007	1898	1887,88	1874	
October	60.4	58.2	59.5	58.0	57.6	57.4	57.2	57.1	57.0	56.5		
	2007	1971	1963	1879	1900	1954	1947	1920	1861,78	1949		
November	48.4	47.9	47.6	47.2	47.0	46.9	46.8	45.2	46.5	46.3		
	2009	1931	2006	1963	1849	2001	1948,75	1896	2005	2003		
December	40.1	39.3	38.5	37.4	37.2	37.0	36.5	36.3	36.0	35.8		
	2006	2001	2011	2012	1982	1824,28	1998	1984	1881	1890,1931		
Annual	53.9	53.75	53.40	53.30	53.1	52.6	52.40	52.39	52.36	52.35	52.31	52.3
	2010	2012	2006	2002	2011	2001	1991	1878	1828	1877	1826	2005

TABLE 30
MIDDLETOWN
1893 – 1908 1951 – 2011
Warmest Months / Years
Corrected for T.O.B

	1	2	3	4	5	6	7	8	9	10	
January	36.8	32.9	32.8	31.0	30.7	30.6	29.8	29.7	29.6	29.3	
	2006	2007	1990	2002	2012	1998	1989	1995	1975	1967	
February	35.3	33.4	32.6	32.3	32.0	31.7	31.5	31.2	30.6	29.8	
	2002	1981	1976	1984	1954	1998	1990	1991	1997	1953	
March	42.5	41.9	41.1	41.0	39.9	39.6	39.5	39.3	38.8	38.1	37.6
	2010	1973	2000	1977	1985	1990	1991	1995	1998	2002	2004
April	51.7	51.4	51.0	50.8	50.0	49.5	49.3	49.2	49.1	48.2	48.0
	2006	1985	1991	1981	1976	1955	2002	1994	1952	2010	2001
May	63.5	62.3	61.4	61.0	60.5	60.1	59.5	59.4	59.0	57.8	57.3
	1991	2004	1975	2010	1980	1986	1955	1982	1972	1999	2001
June	71.3	69.9	69.2	69.1	69.0	68.9	68.7	68.0e	67.6	67.4	67.0
	2005	1957	1999	1984	1976	1987	1983	2010	2001	2006	1989
July	76.9	75.5	75.0	74.8	74.7	74.3	74.0	73.8	73.3.	72.2	
	1955	2006	2005	2002,10	1983	1988	1987	1980,81	1991	1961	
August	74.2	73.5	73.4	73.2	73.8	74.2	70.6	73.0	73.2	73.1	
	1999	2002,05	2001	1988	1995	1980	1981	1973	1983	2003	
September	69.0	69.0	66.6	66.0	65.6	65.1	65.0	64.9	64.8	64.6	
	2005	1961	2007	2010	1959	1980	1983	1998	1985	1971	
October	60.7	57.7	57.6	55.8	54.4	54.1	54.0	53.4	53.2	52.0	
	2007	1971	1963	1990	1968	1995	1955	1959	1970	1998	
November	47.2	46.5	45.7	44.6	44.5	44.4	44.0	43.0	42.5	42.3	
	2006	1975	2005	2001	1994	1963	1999,09	1990	1985	2003	
December	40.3	37.2	36.4	36.1	34.8	34.4	33.6	33.5	32.8	32.1	
	2006	1982,11	1998	2001	1990	1994	1996	1953	1999	1974	
Annual	53.4	51.2	51.0	50.9	50.8	50.5	50.1	49.5	49.4	49.1	
	2006	1998	2002,10	1991	2007	1990	1999	1983	2001	1985	

TABLE 31
MILLBROOK
1942 – 2012
Warmest Months / Years

	1	2	3	4	5	6	7	8	9	10	
January	33.5	33.4	33.0	32.2	32.0	31.2	30.8	30.4	29.9	29.0	
	1998	1990	2002	1995	1950	2012	1949	1953	2008	1947	
February	34.3	34.1	32.4	31.8	31.7	31.5	31.1	30.8	30.0	29.8	
	1998	1997,12	1989	1984	1954	1949	1999	1976	1990	1981	
March	47.4	43.6	43.0	41.7	41.4	40.5	40.3	39.8	39.3	39.2	39.0
	2012	1946	2010	1945	2000	1998	1973	1995	1990	1977	1979
April	52.7	51.3	50.7	50.3	50.6	50.1	49.7	49.3	49.2	49.1	
	2010	2008	1994	1952	2005	1986	2006	1968	2004	1981	
May	62.2	61.7	61.2	60.9	60.8	60.6	60.4	60.3	60.2	59.7	59.5
	2012	2004	1998	1981	2007	1975	1944	1991	1959	2001	1982
June	70.9	70.1	69.7	69.1	69.0	67.9	67.6	67.5	67.4	67.3	
	2005	1957	1943	1999	2008	2001	2007	1973	1949	1995	
July	74.8	74.9	73.9	73.8	73.5	72.9	72.8	72.7	72.5	71.9	71.8
	1955	2002	2006	2011	2012	1957	2008	2005	1988	1987	1952
August	74.3	72.8	72.7	72.6	72.5	71.9	71.5	71.3	71.0	70.6	70.2
	1944	2008	2001	2005	2002	2003	1955	1973	2012	1998	2011
September	67.5	65.9	65.2	65.1	64.2	63.9	63.4	63.2	63.0	62.5	63.1
	1961	2005	2002	2010	1999	1959	2008	1945	1973	1971	2004
October	57.0	54.9	54.6	53.8	53.6	53.5	53.4	53.3	52.9	52.6	
	2007	1971	1947	1946	1963	1990	1961	1949	1959	1975	
November	46.0	45.7	45.5	45.4	44.4	44.3	44.2	43.8	43.3	43.1	
	2001	2006	2011	1999	2005	1963	1948	1975	2003	1966	
December	38.4	36.1	35.9	35.6	35.4	34.7	34.3	33.2	33.0	32.9	
	2006	2011	2001	1998	1996	1990	2012	1973	1971	1953	
Annual	51.7	51.4	51.3	50.9	50.7	50.5	49.8	49.7	49.6	49.5	
	2012	1998	2006	2002	1999	2011	1991	1990	2005,07	1988	

TABLE 32
WARMEST MONTHS / YEARS
MOHONK LAKE
1891 - 2012

	1	2	3	4	5	6	7	8	9	10	11
January	34.4	33.1	32.6	32.2	31.9	31.8	31.7	31.4	31.0	30.7	28.6
	1932	2006	1913	1933	2002	1990	1949	1937	1924	1950	2012
February	33.6	33.2	33.1	32.0	32.3	31.0	30.6	30.4	29.9	29.5	28.5
	2002	2012	1925	1981	1998	1976	1997	1954	1990	1953	1949,91
March	46.7	44.2	44.1	42.3	41.8	41.1	40.0	38.4	38.1	38.0	37.8
	2012	1945	1946	1903	2010	2000	1936	1973	1898	1995	1977
April	54.2	51.8	51.6	51.5	50.5	50.4	50.2	50.1	49.8	49.6	49.3
	2010	2008	1941	1921	2006	2005,09	1968	1945	1910,91	1942	1998,12
May	63.5	62.5	62.0	61.9	61.3	61.0	60.3	59.7	58.8	58.7	
	1991	2004	1944,10	1998,12	1918	1911	1986	1965	1999	1975	
June	71.8	70.2	70.0	69.7	69.6	69.1	69.0	68.8	68.6	68.5	68.1
	2005	2008	1943,99	2001	2010	1925	2007	1895	1923	1949,57	1991
July	76.0	75.9	75.2	75.0	74.4	73.8	73.7	73.4	73.2	72.6	72.0	71.8
	1999	2010	2002,11	2006	2012	2005	1955	1988	2008	1952	1966	1995
August	74.2	73.7	72.3	71.8	71.2	70.7	70.6	70.3	70.1	70.0	69.9
	2001	2005	2002	1998	2012	1995,10	1900	1988,06	1944	1973	1980
September	66.0	65.4	64.7	64.9	64.8	64.2	63.8	63.7	63.6	63.4	63.2
	1961	2005	1998,99	2002	2007	1930	1959	1948	2010	2001	1983
October	58.2	58.0	58.1	56.9	56.0	55.4	55.2	54.7	54.4	54.5	
	1947	1963	2007	1995	1949	1971	1990	1954	1984	2001	
November	46.0	45.5	44.7	44.5	44.9	44.3	44.1	44.5	43.3	43.6	43.1	43.0
	2001	2011	1948	1975	2006	1999	1994	2009	1946,79	2005	1963	2003
December	37.3	35.7	36.2	35.0	34.2	34.1	34.0	33.9	33.5	32.7	32.2	32.1
	2006	1998	2001	2011	1953	1990	1982	1984	1923	2012	1996	1994
Annual	51.3	51.2	50.2	50.3	50.1	50.0	48.8	48.7	48.6	48.2	47.8	
	1998,12	2006	1999,01	2002	2005	2010,11	1990	1991	1949	2000	1995	

TABLE 33
POUGHKEEPSIE
Warmest months /years City and Airport
City 1830 – 1849, 1892 – 1901, 1928 - 1970 · Airport & City (1N) 1949 – 1992
City - 1994 - 2012

	1	2	3	4	5	6	7	8	9	10		
January	35.4	34.8	34.4	34.0	33.9	33.5	32.9	32.6	32.5	32.2		
	1932	1990	1995	1998	2002	1933	1989,12	1967	1937	1949		
February	36.9	36.3	36.1	35.5	34.7	34.3	34.2	34.0	33.9	33.8		
	1997	2012	1998	2002	1981	1984	1960	1953	1976	2006		
March	48.5	45.3	44.5	43.7	43.6	43.3	42.7	42.4	41.6	40.8		
	2012	1945	1946	1977	2010	1973	1936	2004	1995	2000		
April	57.5	53.9	53.6	53.7	53.6	53.5	53.0	51.8	51.7	51.6		
	1941	1960	1945	2008	1977	1969	2010	2006	1955	1991		
May	66.4	65.0	65.1	64.6	64.5	64.4	63.8	63.7	62.7	62.5	61.5	
	1991	1944	1965	1964	1959	2012	2004	1955	1941	2011	2010	
June	75.3	74.3	72.6	72.2	72.0	71.0	70.1	69.8	69.7	69.0		
	1994	1957	1949	2005	1943	1991	2008	2010	1896	2006		
July	78.5	77.9	77.7	77.0	76.4	76.3	76.2	75.8	75.7	75.5	74.9	
	1955	1999	1949	1966	2006	1959	1952	1935,11	2010	2012	2005	
August	77.2	77.0	76.0	75.9	75.3	75.1	75.0	74.5	74.3	73.7		
	1955	1949	1991	1980	1966	2005	1973	1959	1939	1938		
September	71.5	69.4	67.7	67.3	65.5	65.3	65.2	65.1	65.0	64.9		
	1961	2005	1959	2011	1929	2010	2002	1945	2008	2007		
October	59.3	59.0	58.5	58.3	57.4	56.3	56.0	55.9	55.4	55.1		
	1949	1963	2007	1954	1971	1950	1995	1947	2005	1931		
November	49.7	48.8	47.9	47.3	47.2	47.1	46.8	46.0	45.8	45.7		
	1963	1931	2006	2005	1975	1994	1960,11	1948	1999,09	1948		
December	40.3	38.0	38.8	37.8	37.2	37.1	37.0	36.3	36.0	35.4		
	2006	1990	1999	2011	1953	2001	1998	1957	2012	1965		
Annual	53.8	53.7	53.2	53.1	53.0	52.1	51.8	51.6	51.2	50.8		
	2006	2012	1953	1949	1990	2005,11	2010	1998	1941	1995		

TABLE 34
WARMEST MONTHS
Corrected for Time of Observation and Urbanization
ALBANY CITY/ AIRPORT
1820 - 2012

	1	2	3	4	5	6	7	8	9	10	
January	32.9	31.6	31.5	31.3	30.0	29.9	29.8	29.7	29.1	29.0	28.9
	1932	1913	1933,06	1990	2006	2002	1950,95	1937	1880	1906	1949
February	33.0	32.1	31.6	31.0	30.9	30.3	30.2	30.1	30.0	29.5	28.9
	1828	2012	1981	1840,42	1984	1998	2002	1954	1976	1857	1997
March	45.9	42.8	42.6	41.8	41.6	41.0	40.7	40.4	39.2	38.8	38.7
	2012	1945	1946	2010	1871	1903	1973	1921	1878	1977	2000
April	52.6	52.0	51.9	51.7	51.1	51.0	50.9	50.4	50.2	50.0	
	1878	1830	2008,10	1941	1952	1921	1844	1915	2005	1931	
May	65.0	64.0	63.5	63.2	63.1	63.0	62.9	62.0	61.7	61.5	61.2
	1826	1887	1944	1911	2012	1860	1998	1991,11	1864	2004	2010
June	73.0	72.5	72.0	71.2	71.5	71.1	71.0	70.6	70.5	70.1	
	1858	1828	1831	1943	1825	2005	1949	1895	1838	1930	
July	77.4	76.5	76.0	75.6	75.5	74.9	74.8	74.5	74.2	74.0	
	1868	1820	1825	1854	1887	2006,10,11	2012	1834	1921	1935	
August	74.0	73.2	73.1	73.0	72.7	72.6	72.5	72.4	72.2	72.1	
	1872	1937	1939	1845	1900	1877	1947	1938	2012	2005	
September	67.7	67.3	66.0	65.2	65.0	64.6	64.3	64.1	64.0	63.9	63.8
	1881	1961	1846	1865	2005	2010	2007	1900	1841	1891	2002
October	56.8	55.6	55.2	54.1	53.9	53.6	53.5	53.0	52.2	52.3	
	2007	1947	1879	1900	1920,63	1949	1954	1946,71	1995	1961	
November	47.6	46.7	45.0	44.4	44.0	43.8	43.6	43.5	43.4	42.9	
	2006	1991	1830	2010,11	1849	1948	2001	1931	2009	1999	
December	35.2	35.0	34.0	33.5	33.3	33.0	32.7	32.3	32.2	32.1	
	2006	1829	2011	1848	2012	1852	2001	1984,96	1998	1990	
Annual	51.69	50.46	50.2	49.70	49.67	49.6	49.00	48.89	48.81	48.48	48.41
	2012	2006	2010	130	1828	2011	1998	1953	1949,90	1870	1931

TABLE 34A
ALBANY CITY & AIRPORT
1890 – 2012
Warmest Months / Years • Corrected for T. O. B (1890 – 1938)

	1	2	3	4	5	6	7	8	9	10	11
January	32.9	31.6	31.5	31.3	30.0	29.9	29.8	29.7	29.0	28.9	28.6
	1932	1913	1933	1990	2006	2002	1950,95	1937	1906	1949	2012
February	32.1	31.8	31.6	30.9	30.2	30.1	30.0	29.9	29.8	29.3	28.9
	2012	1998	1981	1984	2002	1954	1991	1976	1925	1949	1953
March	45.9	42.8	42.6	41.7	41.0	40.7	40.5	40.4	40.2	40.0	38.8
	2012	1945	1946	2010	1903	1973	1936	1898	1921	1995	1998
April	51.9	51.7	51.2	51.1	51.0	50.8	50.4	50.2	50.0	49.6	
	2008,2010	1941	1991	1968	1921	1945	1915	2005	1931	2006	
May	61.3	63.5	63.2	62.9	62.0	61.5	61.2	61.1	61.0	59.6	59.5
	1986	1944	1911	1998	1991	2004	2010	1908	1918	1965	1988
June	72.0	71.2	71.1	71.0	70.8	70.4	70.1	69.7	69.0	68.4	
	1895	1943	2005	1949	1957	2008	1930	1999	1991	2001	
July	76.3	75.0	74.9	74.8	74.2	74.1	74.0	73.9	73.8	73.4	
	1955	1988	2006,10,11	1952,12	1999	1921	1995	1935	2005	2002	
August	73.7	73.2	73.1	72.9	72.7	72.5	72.4	72.2	72.1	72.0	71.1
	2001	1937	1939	2002	1900	1945	1938	2012	2005	2010	1998
September	67.3	65.8	65.0	64.8	64.3	64.1	63.9	63.8	63.3	63.2	
	1961	1930	2005	1999	2007	1900	1891	2002	1985	1998	
October	56.8	55.6	55.5	54.7	54.2	54.1	53.8	53.6	53.5	53.4	
	2007	1947	1963	1971	1946	1900	1984	1949	1954	1920	
November	47.6	46.7	45.5	44.4	43.8	43.6	43.5	43.3	42.9	42.6	
	2006	1991	1975	2010,11	1948	2001	1931	1994	1999	1946	
December	35.2	34.0	33.8	33.7	33.5	33.3	32.7	32.2	32.1	31.6	
	2006	2011	1984,96	1982	1953	2012	2001	1998	1990	1994	
Annual	51.69	50.46	50.2	49.9	49.6	49.4	49.1	49.00	48.89	48.81	48.41
	2012	2006	2010	2011	2002	1999,01	1908	1998	1953	1949,80	1931

TABLE 35
ALCOVE DAM
1949 – 2012
Warmest Months / Years · Corrected for Time of Observation

	1	2	3	4	5	6	7	8	9	10	11
January	31.1	30.3	30.2	30.1	29.8e	29.0	28.6	27.8	27.6	26.3	
	1990	2002	2006	1995	1950	1949	2007	1998	2012	1953	
February	31.8	31.6	30.9	30.4	30.3	30.2	29.9e	29.7	29.2	28.0	
	1984	2012	1954	1998	2002	1981	1991	1949	1997	1976	
March	43.5	40.4	37.9	36.8	36.7	36.5	35.9	35.8	35.7e	35.3	35.2
	2012	2010	2000	1977	1995	1973	1998	1991	1951	1985	1949
April	50.7	48.1	48.0	47.8	47.7	47.5	46.8	46.7	46.4	46.3	
	2010	2009	1991	2002	1955	2006	1952	1981	1994	1985	
May	60.5	60.0	59.6	59.2	59.1	59.0	58.4	57.9	57.7	57.3	56.1
	2012	2010	1991	2004	1998	1975	1986	2001	1985	1980	1999
June	68.9	68.5	67.0	66.9	66.2	65.6	5.2	64.8	64.7	64.4	64.3
	2005	1957	1984	1983	1999,10	1976	2001	1987	2006	2011	2012
July	75.6	73.4	72.8	72.1	72.0e	71.4	71.0	70.9	70.8	70.7	70.6
	1949	1955	2010	2012	1987	2006	1988	2005	1981	1999	1995
August	73.5	72.0	71.4	71.0	70.7	70.3	70.0e	69.9	69.4	69.3	
	1973	1955	1999	2005	2002	2003	1990	2001	1980	1984	
September	67.6	64.1	64.2	63.1	62.6	62.4	62.1	62.0	61.6	61.5	61.4
	1961	1959	2005	2011	2010	1999	1971	1983	2007	1980	1998
October	54.5	54.2	53.7e	51.8	51.4	51.3	50.9	50.6	49.9	49.6	
	2007	1963	1999	1971	1990	1984	1959	1955	1970	1968	
November	44.9e	44.3	43.8	43.6	43.5	43.3	42.7	42.3	42.7	41.2	40.9
	1948	2006	1999	2010	1994	2009	1963	1975	2001	2003	2005
Decembe	35.3	33.9	33.5	33.1	32.9	32.8	32.7	31.9	31.3	29.5	
	2006	2011	2001	1996,12	1998	1982	1990	1953	1994	1999	
Annual	49.8	49.5	48.8	48.7	48.2e	47.9	47.8	47.6	47.5	7.3	47.0
	2012	2010	2006	1998	1990	2002	1991,11	1999	2001	1953	1955

TABLE 36
GLENS FALLS
1893 – 1933 1943 – 1947 (Corrected for TOB) • Fire Station 1948 – 1956 Farm 1957 - 2007 • Airport 1948 - 2012
Warmest Months / Years

	1	2	3	4	5	6	7	8	9	10	
January	29.4	29.3	28.8	28.4	28.3	28.0	27.8	27.4	27.2	26.7	26.0
	1950	1990	1913	1995	1933	1949	2006	2002	1998	1906	2012
February	32.2	30.4	30.0	29.3	29.1	28.8	27.5	27.4	26.4	26.2	26.0
	1981	1984	1998	1954	1949	2012	2002	1991	1976	2010	1997
March	42.4	41.8	41.1	40.0	38.9	38.8	38.7	38.1	38.0	37.7	36.7
	1946	2012	1945	1903	2010	1921	1973	1977	1949	2000	1995
April	51.3	51.0	50.2	50.0	49.8	49.2	49.0	48.6	48.0	47.9	47.8
	2008	1946	1921	1915	1945	2010	1991	1968	2006	1910	1998
May	63.0	62.0	61.7	61.1	60.9	60.6	60.3	60.4	59.8	59.6	58.7
	1911	1998	2011	2010	1991	2012	1918	1964	1975	1965	2004
June	70.9	70.8	69.0	68.3	68.2	67.9	67.8	67.6	67.0	66.6	
	1949	2005	1895	1919,99	2008	2010	1930,67	1994	1976	2001	
July	76.3	76.1	75.4	74.8	73.6	73.1	73.0	72.4	71.8	71.6	71.2
	1949	1955	1952,11	2010	1921	2006	1988	2005	1999,11	1901	2012
August	75.8	71.8	71.7	71.6	71.3	70.5	70.2	69.4	69.3	69.2	
	1955	2005	1959	1973	2001	1988	1980	2002,03	2012	1995	
September	67.5	65.0	63.5	63.4	63.0	62.5	62.3	61.3	61.2	59.6	
	1961	2005	1945	1999,11	2007	2002	1947	2010	1998	1900	
October	57.2	6.4	55.8	54.4	54.1	53.9	52.7	52.6	52.5	51.1	
	1947	1949	1954	1963	2007	1971	1995	1920	1900	1984	
November	45.8	43.7	43.0	42.9	41.9	41.8	41.1	41.0	40.9	40.5	
	1948	1963	2006	1999	1975	2011	2009	1946	1994	2001	
December	33.9	33.8	32.3	32.0	31.7	31.5	31.4	31.3	31.1	30.8	30.2
	2006	1953	1996	1957	1998	1923	2011	1984	2001	1982	2012
Annual	49.4	49.0	48.9	48.4	48.7	48.1	47.7	47.6	47.1	46.8	46.5
	1949	1953	1998	2010	2012	2006	1990	1999	2011	2005	2002

TABLE 37
SARATOGA SPRINGS
Warmest Months/ Years
1941 - 2012

	1	2	3	4	5	6	7	8	9	10	11
January	28.3	28.2	27.2	26.5	26.2	26.0	25.5e	25.0	24.6	24.3	
	1990	2006	2002	2012	1995	1950	1953	1989	1998	1949	
February	31.9	28.7	27.6	27.0	26.7	26.4	25.7	24.3	23.9	23.4	22.9
	2012	1981	1984	2002	1998	2010	1991	1997	1999	1949	2005
March	46.0	41.3	40.8	40.5	38.0	36.8	35.4	35.0	34.7	34.4	34.2
	2012	2010	1945	1946	1977	2000	1973	1998	1995	1985	1986
April	51.6	48.8e	48.6	48.1	47.7	47.5	47.3	47.1	47.0	46.6	46.2
	2010	1941	1991	2008	1986	2005	1945	1968	1998	2002	1985
May	62.4	61.6	61.1	59.3	59.0	58.1	57.2	57.0	56.8	56.6	56.4
	2012	1991	2010	1944	2004	1986	1942	1965	1999	1964	1959
June	70.2	70.0	68.6	68.1	67.9	67.8	67.0	66.6	66.5	66.3	65.9
	2005	1957	1943	1999	2010	1949	2012	2001	1976	2008	1991
July	74.5	74.3	73.3	73.0	72.5	72.4	72.2e	71.4	71.2e	70.2	69.2
	2010	2012	2011	1995	1999	2005	1955	2002	1952	2006	1944
August	71.8	71.3	71.1	70.5	70.1	70.0	69.8	69.7	69.6	68.5	
	2005	2001	2012	1959	2010	1955	1980	1988	2002	1995	
September	66.8	64.7	63.6	63.2	62.8	62.3	60.7	60.5	60.2	60.1	59.7
	2011	1961	2005	2010	2002	1999	2007	1998	1948	1959	2001
October	54.5	53.6	51.6e	51.5	51.3	51.1	50.6	50.4	50.0	49.7	
	1947	2007	1954	1949	1963	1971	1995	1990	2001	1984	
November	42.6	41.7	41.0	40.8	40.0	39.7	39.4	39.3	39.0e	38.8	
	2009,11	1999	1948	2001	1963	1979	1994	2005	1975	2003	
December	37.0	32.4	31.9	31.7	31.5	31.3	31.0e	30.8	30.7	30.3	28.7
	2012	2006	2001	2011	1998	1996	1953	1984	1982	1990	1994
Annual	49.4	48.6	48.2	47.7	47.1	47.0	46.9	46.8	46.7	46.6	46.4
	2010	2011	2012	1998	1990	1991	2001	2006	2002	1999	2005

TABLE 38
TROY (Lock & Dam)
Warmest Months
1913 - 1935 1956 - 2012

	1	2	3	4	5	6	7	8	9	10	
January	35.3	35.2	34.4	33.7	33.0	32.3	30.4	30.3	29.9	29.4	28.8
	1933	1932	1990	2002	2006	1995	1998	2012	1967	1989	1993
February	34.3	33.6	33.5	33.0	32.5	32.2	32.0	31.4	30.5	28.7	
	2012	1984	2002	1981	1991	1998	1997	1976	1990	2010	
March	45.5	42.5	41.3	39.9	39.8	39.4	39.6	39.0	38.5	38.2	
	2012	2010	2000	1995	1973	1921	1998	1977	1990	1983	
April	52.4	52.2	51.5	51.2	50.6	50.5	50.3	50.0	49.8	49.4	49.3
	2010	2008	1986	1987	1968	2005	2002	1998	1921	2009	2006
May	64.5	63.5	63.0	62.4	61.7	61.2	60.8	60.7	60.6	60.3	60.0
	1998	2012	1991	2004	1918	1999	1975	1965	1964	1977	1989
June	73.0	72.3	71.4	70.7	70.6	70.5	70.4	70.3	70.1	69.8	
	2005	1933	1999	1957,76	1991	1994	2008	1967	1930	1919	
July	76.4	76.2	76.1	76.0	75.5	75.3	75.0	75.1	74.4	73.1	
	1995	1999	1921,10	2006,12	1988	2011	1935	2002	2005	1966	
August	74.9	74.5	74.4	74.3	74.1	74.0	73.6	73.3	72.3	72.2	72.1
	1973	1988	2005	2002	2003	2001	2012	1998	2010	1959	1995
September	68.3	68.0	67.5	67.3	67.2	67.0	66.7	66.4	66.0	65.6	
	1961	2005	2002	1999	2011	2010	2007	1971	2004	1983	
October	58.5	56.9	56.7	56.5	55.4	55.0	53.9	53.0	52.6	49.9	
	2007	1971	1995	1984	1990	2001	1920	1963	1961	1931	
November	47.6	47.5	46.9	46.6	46.1	46.0	45.9	45.8	44.5	44.4	
	2001	1999	1991	1975	2011	1979	2006	1994	2005	2003	
December	38.3	37.4	37.0	36.4	36.1	36.0	35.9	35.3	34.2	33.6	
	2006	2001	1998	1996	1982	1990	2011	1984	2012	1957	
Annual	52.9	52.7	51.7	51.6	51.5	51.3	51.2	51.1	50.8	50.4	50.3
	2012	1998	2006	1990	1999	2001	2002	1991	2011	2005	2010

TABLE 39
LITTLE FALLS 1 – CITY RESERVOIR
1897 – 1899; 1901 – 2012
Warmest Months / Years   missing 1999, 2000 (Used Utica) Corrected for Time of Observation

	1	2	3	4	5	6	7	8	9	10	
January	31.3	30.8	29.6	29.5	27.7	27.6	25.9	25.3	25.2	24.6	
	1990	1932	2002	1933	2006	1950	1995	1998	2012	1916	
February	29.8	29.7e	29.2	29.1	29.0	28.2	27.8	26.5	26.4	26.3	
	1981	1999	1984	1954	2012	2002	1998	1925	1976	1949	
March	43.4	41.5	41.3	40.0e	38.6	37.7	37.5	36.3	36.1	34.5	34.3
	2012	1946	1945	2000	1903	2010	1921	1936	1977	1942	1998
April	51.3	50.4	49.8	49.0	48.9	48.6e	47.9	47.4	47.3	46.0	45.9
	1915	1921	1941	1945	1942	2010	1986	1952	1991	2005	2009
May	61.6	61.0	60.6	60.5	60.1	60.0	59.7	58.7	58.6	58.5	
	1998	1903	2012	1918	1975	1944	1991	1936	1986,10	2004	
June	69.1	69.0	68.5e	68.4	67.8	67.0	66.9	66.8	66.7	66.6	
	1999	1949	2005	1943	1919	1930	1938	1957	1933	1976	
July	74.0	73.8	73.3	72.3	72.2	72.1	72.0	71.8	71.4	71.2	
	1999	1921	1955	1949,11	1935	1916	2006	1901	1964	2010	
August	71.9	71.4	71.1	70.9	70.8	70.6	70.0e	69.6	69.2	68.5	67.6
	1937	1939	1955	2001	1973	1938	1988	2003	1959	2010	1980
September	65.4	64.6	64.5	64.4	63.8	63.6e	63.0	62.5	61.8	61.6	60.5
	1999	1915	1961	2005	2007,11	2004	1998	2010	1945	1931	2003
October	56.0	55.8	53.2	53.1	52.5	52.2	52.0	51.3	50.7	50.6	
	1947	1971	2007	1920	1946	1954	1995	2001	1915	1961	
November	44.0	43.0	42.4	42.3	42.0	41.7	41.5	41.1	40.9	40.1	
	1999	1948	2006	2001	2009	1931	1963	1979	1975	1982	
December	34.9	34.0	32.4	31.6	31.2	31.0	30.8	30.7	30.4	30.3	
	2006	1996	1923	2011	2012	1953	1998	1957	1984	1911	
Annual	49.2	49.0e	48.2	47.6	47.3	47.2	47.1	47.0	46.9	46.8	
	2012	2005	1998	2011	1990,10	2006	1949	1953	1921	1938	

TABLE 40
WARMEST MONTHS
Corrected for Time of Observation
INDIAN LAKE
1899 - 2012

	1	2	3	4	5	6	7	8	9	10	
January	27.3	27.2	27.1	26.8	26.5	26.1	25.7	25.1	24.6	24.2	
	1937	1990	1932	1950	1949	1933	2006	1913	2002	1995	
February	27.6	26.5	26.4	26.3	25.6	25.1	24.6	24.3	24.1	23.9	23.1
	1986	1953	1984	1998	1981	1954	1925	2012	1949	2002	1991
March	36.0	35.8	34.8	34.6	34.5	33.8	33.3	32.8	31.9	31.3	
	1946	2012	2010	1903	1921	1945	2000	1973	1977	1995	
April	47.3	45.2	44.0	43.8	43.7	43.6	43.0	42.7	42.5	42.4	
	1915	1921	1987	2010	1986	1991	1910,41	1945,55	1998	2008	
May	59.2	58.0e	57.5	56.4	56.2	55.4	54.9	54.8	54.7	54.3	54.1
	1911	2012	1908	1944	1918	1998	1986	1955,91	1975	1988	2004
June	65.3	64.1	63.7	63.0	63.2	62.6	62.2	62.1	62.0	61.9	
	2005	1949	1919	1925	1957	1901	1943	1976	1930	1999	
July	73.5	70.0	68.3	68.0	67.4	67.3	67.2	66.5	66.2	66.1	66.0
	1901	1921	1955	1901,16	1908	2011	1988	1949	2005	2006,10	1999
August	66.8	66.4	66.0	65.9	65.6	65.5	65.3	65.0	64.7	64.3	
	1955	1947	1937	1959	1988	2001	2005	1900	1944	1939,11	
September	62.5	60.0	59.9	59.5	58.5	58.4	58.2	58.1	58.0	57.9	
	1961	2011	2002	1900	1999	2005	2007	1959	2004	1931	
October	51.1	50.9	50.5	50.0	49.1	49.0	48.7	48.2	47.8	47.6	
	1947	1971	1900	2007	1990	1920,95	1949	1954	1984	1914	
November	40.2	39.4	39.2	38.6	38.4	38.3	38.2	37.8	37.5	36.9	
	1948	2006	2011	1999	1994	1963,79	1953,09	1931	1902	1908	
December	30.5	29.3	29.2	29.1	28.4	27.8	27.7	27.5	27.4	27.3	
	2006	1984	2001	1998	1923	1957,96	2012	1982	1990	1911	
Annual	44.0	43.8	43.7	43.2	43.1	43.0	42.9	42.8	42.6	42.4	42.3
	2012	1998	1990	2011	1921,53	2006	2010	1949	1991	2002	2001

TABLE 41
EXTREME DAILY MAXIMUM TEMPERATURES WITH MONTH OF OCCURRENCE

	Jan	Feb	Mar	Apr	May	Jun	Jul	Aug	Sep	Oct	Nov	Dec
HUDSON VALLEY DIVISION												
Albany City & AP	71	68	89	93	97	100	104	102	100	91	82	71
(1895 -	2007	1997	1998	1941	1911	1933	1911	1918	1953+	1941	1950	1984
Alcove Dam	62	70	86	91	96	96	98	102	96	89	81	69
(1949 -	2008	1985	1984	2002	1964	2008+	1949	1955	1953	1963	1982	2001+
Glens Falls	64	66	86	92	98	100	103	102	101	88	82	69
F&AP (1896-	1995	1957	1998	1915	1911	1952	1949	1949	1953	1947	1950	1982
Middletown	66	73	85	92	93	97	101	97	100	88	78	71
(1952 -	2007	1954	1998	1976	1962	1952	1995	2002+	1953	1963	1982+	1998
Mohonk Lake	65	70	84	91	92	96	98	101	95	85	78	66
(1896-	1950+	1985	1998	2002+	2000	2005	2006	2001	1953	1941	1950	2001
Peekskill	72	71	84	95	96	102	104	102	101	91	83	72
(1946 -	1950+	1985+	1945	2002	1996	2005	1937	1948	1953	1941	1950	2001
Poughkeepsie	68	73	86	94	102	102	107	104	103	91	84	72
C& AP (1928 -	1932	1997	1998	1976	1962	1933	1966	1949+	1953	1949+	1950	1998
Saratoga Springs	63	64	88	92	92	99	97	102	94	89	82	67
(1941 -	1932	1981	1998	1990	1962	1964	1949+	1947	1973+	1947	1982	1998+
Troy	70	71	83	91	97	101	108	102	98	99	83	69
(1913 -	1932	1930	1998	2002	1914	1920	1926	1918	1933+	1927	1975	1934
West Point	71	72	86	95	97	102	106	105	105	92	82	72
(1890 -	1950+	1930	1945	1976	1962+	1952+	1926	1931	1953	1941	1950	1998
Yorktown	71	73	88	95	96	100	103	102	103	92	81	70
(1888 -	2007	1985+	1945	2002	1962	1950	1936	1948	1953	1941	1950	1984

MOHAWK VALLEY DIVISION

	Jan	Feb	Mar	Apr	May	Jun	Jul	Aug	Sep	Oct	Nov	Dec
Little Falls	62	61	82	90	95	96	98	98	95	87	78	70
CR (1897 -	2008	2002	1945	1962	1945	1948	1936	1916	1983	1900	1982	1901
Gloversville	68	62	83	92	96	97	100	99	99	87	77	66
(1893 -2005)	1950	1997	1998	1962	1911	1942	1911	1916	1953	1949	1950	2001
Utica	65	64	87	91	98	100	100	100	98	89	79	70
(1893 – 2006)	1995	1976	1986	1990	1914	1947	1953+	1948+	1953	1941	1982+	1966

NORTHERN PLATEAU DIVISION

	Jan	Feb	Mar	Apr	May	Jun	Jul	Aug	Sep	Oct	Nov	Dec
Indian Lake	62	61	82	86	95	98	103	100	92	87	75	64
(1900 -	2008	2002	1945	1915	1911	1907	1911	1916	1945	1905	1950	2001
Newcomb	59	59	76	88	91	95	94	94	91	85	74	62
(1940 -	2008+	1994	1990+	1990	1966	1944	1953+	1988+	2002	1950+	1950	2001

+ = more than one date with the same temperature

HEATING DEGREE DAYS

Heating degree days are used as a measure in determining the amount of fuel needed to heat homes when the average temperature of the day (high plus low divided by two) falls below 65 degrees Fahrenheit. For example, a day with an average temperature of 60 degrees will have 5 degree days and most homes or businesses will require heat. Within most of the Hudson Valley watershed there are generally few or no heating degree days during the summer months of June, July and August.

The use of heating degree days is therefore another method to determine temperature trends. For a given heating season, the greater the number of heating degree days the colder the season and conversely the lower the number the warmer the season.

Fifty years of daily temperature records for sixteen long-term weather stations has shown a clear pattern of warming from the 1980s through the first decade of the 21st century. (Table 42). Of the twelve stations listed, all but one show that the decade 2001 – 2010 has the fewest Degree Days of the decades shown indicating the warmest heating season.

HEATING DEGREE DAY TRENDS

In an examination of twelve weather stations in the Hudson Valley Climate Division, one Mohawk Valley station and one Northern Plateau station (Table 42) reveals a distinct reduction in the number of annual degree days over a fifty year period for all stations with unbroken records. The average reduction for all twelve was 616 degree days, with Saratoga Springs showing the greatest reduction (1028) and Glens Falls Airport the least (328). This reduction indicates a significant warming during the normal heating season from October through May in the Hudson Valley watershed.

TABLE 42
HEATING DEGREE DAYS
Decadal Averages

	Albany AP	Alcove Dam	Glens Falls Airport	Middletown	Mohonk Lake	Poughkeepsie	
1961- 70	7103	7592	7628	6159	6727	6534	
1971 -80	6968	7484	7669	5885	6594	6267	
1981 -90	6674	7177	7324	5696	6274	6331	
1991- 00	6677	7274	7342	5741	6266	6255	
2001- 10	6414	6958	7300	5811	5866	6025	

	Saratoga Springs	Troy	West Point	Yorktown Heights	Little Falls	Indian Lake
1961- 70	7395	6857	5868	6352	7726	9447
1971- 80	7265	6658	5536	6097	7646	9306
1981- 90	6767	6457	5270	5981	7616	8846
1991- 00	6739	6446	5540	5959	7745	8978
2001- 10	6367	6126	5519	5619	7377	8608

THE FROST FREE SEASON

The records at fifteen long term stations reveal a consistent if erratic higher trend in the length of the frost free season over the last 50 years. At ten of fifteen stations, the decade of 2001through 2010 shows the greatest average number of days between the last and first frost equal or less than 32 degrees Fahrenheit. These have occurred at stations in all three of the Hudson Valley climate divisions (Table 43). At all but two stations (Glens Falls Airport & Troy) the three decades since 1981 show the average greatest number of frost free days when compared with their previous decades of daily temperature observations.

As can be expected, the greatest number of frost free days occurs in the southern part of the Hudson valley, namely at West Point and Yorktown Heights where over 200 or more days are revealed, while Indian Lake in the Adirondacks shows the fewest number. At that location, a last frost can occur in the latter part of May and first frost in mid September. To the south, the last three decades have had many of their last frost days in the last week of March and their first in the first week of November.

Table 44 also shows most of the greatest number of frost free days at fourteen stations for specific years with most of these occurring since 1981.

At the Yorktown Heights Cooperative station the average date of the last frost for the thirty year periods of 1971 -2000 and 1981 -2010 periods was April 13. The average date of the first frost for the 1971 -2000 period was October 20 and was October 21 for the 1981-2010 period.

TABLE 43
AVERAGE NUMBER OF DAYS BETWEEN
LAST AND FIRST FROST =< 32 F
By decade

	West Point	Troy	Mohonk Lake	Indian Lake	Little Falls CR
1921- 30	188		172	86	146
1931- 40	195		179	94	149
1941- 50	188		177	98	145
1951- 60	189		189	98	144
1961- 70	191	166	174	90	132
1971- 80	183	172	179	96	141
1981- 90	202	184	179	105	138
1991- 00	200	186	182	120	147
2001- 10	197	177	180	126	153

	Yorktown Heights	Albany Airport	Alcove Dam	Glens Falls Farm	Glens Falls Airport
1921- 30					
1931- 40					
1941- 50					
1951- 60	149#	141	133	147	146
1961- 70	158	144	132	130	137
1971- 80	180	147	120	132	143
1981- 90	182	156	144	148	149
1991- 00	184	156	153	143	144
2001- 10	204	163	157	149	140

Carmel

	Gloversville	Utica Airport	Middletown	Poughkeepsie	Saratoga Springs
1914- 20	135				
1921- 30	145				
1931- 40	149	146		179	
1941- 50	135	145		171	
1951- 60	143	162		169	136
1961- 70	127	156	172	168	117
1971- 80	150	164	186	167	145
1981- 90	149	154	186	155	141
1991- 00	146	160	182	163	144
2001- 10	150*	158*	193*	164*	159

* = estimated, 1 or more years missing

BELOW ZERO TEMPERATURES

There has always been a fascination with "below zero" as representing the epitome of cold especially for those living in the urbanized eastern U.S.

Within the Hudson Valley watershed, zero and below temperatures have been a common winter season event. Even at present in the Northern Plateau climate division region below zero minimum temperatures can be expected every winter season. Other weather stations besides the long-term station at Indian Lake show this fact. This is nearly so for the Mohawk Valley climate division stations at Little Falls, Gloversville and Utica, where except for a few very warm winters since the late 20th century all have all recorded zero or below daily minimums. Within the Hudson Valley climate division a different pattern is seen. For the stations at about the same latitude as those of the Mohawk Valley such as Saratoga Springs and Glens Falls, below zero minimums are also quite the norm till the 1990s.

For the mid and lower Hudson valley south of Albany there are now fewer winter seasons with zero or below minimums. In 1990 and 1991 Middletown and West Point had no zero or below lows, in 1992 and 1998 there were no such lows at Yorktown, West Point and Troy. In the lower Hudson valley at Yorktown, West Point and Mohonk Lake there were no such lows in 1999, 2001, 2002, 2005, 2006, and 2008.

From 1896 through 1990 at the Carmel /Yorktown Heights combined cooperative stations there have been only ten years (1909, 1911, 1937, 1946, 1949, 1953, 1956, 1966, 1969, & 1989) when the lowest daily minimum temperature did not drop below zero. Since 1991 through the 2011 – 2012 winter season, there have been eleven such years with no below zero minimums.

In summary, since the start of the 20thst century at the Carmel/Yorktown station there have been 20 years with no below zero minimums. Ten of these years have occurred since 1991. In the northern reaches of the Hudson Valley watershed the number of such days in the winter season has decreased in the period (1991– 2012).

Climatological winter includes the three coldest months of the year, December, January and February. However there have been a few days in November and March with zero and below minimums. These have occurred mainly in the northern region of the Hudson valley watershed. At Indian Lake since 1900 there have been six Novembers that have recorded such low marks. These were in 1904, 1933, 1936, 1951,1956 and 1962. In November 1933 and 1951 below zero lows were also set at Glens Falls.

Below zero marks were much more frequent in March than November with lows set in the north at the six official Weather Bureau stations at Indian Lake, Glens Falls, Gloversville, Saratoga Springs, Little Falls, and Albany in 1905, 1906, 1913, 1914, 1916, 1923, 1934, 1941, 1943, 1950, 1956, 1962, and 1967. In the lower and mid Hudson Valley Carmel, West Point, Poughkeepsie and Mohonk Lake had zero and below in 1906, 1916, 1923, 1934, 1941, 1943, 1950, 1962, and 1967. It should be noted that there have been no zero or below daily minimums in November and March in the entire watershed since 1967.

EXTREME RECORD LOW TEMPERATURES

Another way of indicating the warming that has pervaded the Hudson Valley watershed is to review the frequency of new daily record minimum temperatures at seven representative stations with over 100 years of readings For a 30 year period assuming no temperature trend, three to four new lows would be expected by chance. As is shown below there are fewer in the 1981 – 2010 period for seven stations for most of the months of the year.

FREQUENCY OF NEW RECORD DAILY MINIMUM TEMPERATURES

INDIAN LAKE - 1900 – 2012

January - one after 1981 (1994)
February - two after 1979 (1993) & (2008 leap year)
March – two after 1980 (1984) & (2007)
April - none after 1977
May – none after 1966
June – none after 1979
July – none after 1962
August – none after 1982
September – none after 1965
October – none after 1965
November – two after 1972 (1991 Pinatubo)
December – none after 1980

LITTLE FALLS CITY RESERVOIR 1897 – 2010

January – three after 1981 (1994) (1996) (2004)
February – one after 1980 (1995)
March – five after 1975 (1984) (1988) (1993) (2007)2X
April - four after 1980 (1992 Pinatubo) (1995)2X (1997)
May – four after 1978 (1983) (1985) (1986) (2002)
June - three after 1980 (1985) (1986) (1988)
July – One after 1979 (1985)
August – two after 1982 (1984) (2004)
September - five after 1979 (1984) (1987) (1990) (1991 Pinatubo) (1992)
October - three after 1978 (1984) (1993) (2002)
November – three after 1983 (1984) (1986) (1987)
December – three after 1980 (1988) (1989) (2005)

GLOVERSVILLE 1895 – 2006

January – nine after 1982 (1984) (1994)5X (1996)2X (2004)
February – none after 1980
March – six after 1980 (1984) (1993)4X (1996)
April – two after 1976 (1997)2X
May – two after 1983 (2002) (2006)
June – one after 1979 (1997)
July - none after 1979
August - two after 1972 (1994) (1997)
September – one after 1964 (2006)
October – three after 1984 (1993) (1996)
November – three after 1971 (2000) (2002) (2005)
December two after 1980 (1988)2X (1989)2X

ALBANY CITY & AIRPORT 1874 - 2012

January – three after 1984 (1994) (1996) (2005)
February – two after 1979 (1987) (1995)
March – four after 1980 (1984)2X (1993) (1996)
April – three after 1981 (1982)3X
May – three after 1981 (1986) (2002)2X
June – two after 1980 (1985) (1993)
July – two after 1982 (1985) (1997)
August – four after 1982 (1987) (1988) (1994) 2X
September –one after 1978 (1991 Pinatubo)
October – three after 1981 (1984) (1988) (1993)
November – one after 1981 (2000)
December – four after 1973 (1988)3X (1989)

MOHONK LAKE 1896 – 2012

January – four after 1982
February – one after 1979 (1993)
March – eight after 1980 (1984)2X (1986) (1989) (2003) (2007) 2X
April – two after 1972 (1979) (1980)
May – three after 1976 (2002)2X
June – three after 1982 (1985) (1986) (1988)
July – none after 1978
August – one after 1981 (1989)
September – one after 1974 (1994
October – three after 1974 (1996) (2000) (2003)
November - six after 1976 (1986)2X (1987)2X (1993) (1996)
December – three after 1980 (1988) (1989)2X

WEST POINT 1896 – 2012

January – six after 1981 (1983) (1985) (1994)2X (1996)2X
February - one after 1979 (2007)
March – five after 1975 (1984) (1986) (1989) (1993) (2007)
April – one after 1981 (1993)
May – three after 1978 (1983) (2002)2x
June - five after 1980 (1988) (1996) (1997)2X (1997)
July – three after 1982 (1996) (1997)2X
August – two after 1979 (1983) (1986)
September – four after 1979 (1984) (1985) (1994)2X
October – three after 1982 (1984) (1987) (1993)
November – six after 1976 (1986)2X (1987)3X (2004)
December – two after 1980 (1989)2X

CARMEL/YORKTOWN HEIGHTS 1888 – 2012

January – one after 1981 (1994)
February – none after 1979
March – one after 1980 (1993)
April – none after 1982
May – none after 1976
June – none after 1980
July – none after 1973
August – two after 1968 (1982)2X
September – none after 1964
October – one after 1978 (1993)
November – none after 1978
December – none after 1982

PLANT HARDINESS ZONES

Plant Hardiness zones are used to determine what is the normal coldest in a winter season. Different vegetation can exist and propagate in the different zones. In the Hudson Valley watershed this relates to zero and below temperatures. Within the watershed there is a wide range of normal winter minimum temperatures which can range from near zero in the south to 30 below zero in the Northern Plateau Adirondack region. Hardiness zones are given in five degree intervals.   In the south it can range from zero to five above (7a). In the most northern part of the watershed region, the range is from 30 to 25 below zero (4b). Between these are mid- Hudson valley (5b) 15 to 10 below zero; and the Mohawk valley (5a) 20 to 15 below zero. The base line used by the U.S. Department of Agriculture was the period from 1974 to 1986. These ranges were changed in January 2012 to reflect a 5 degree warming using the period of 1976 – 2005 as is indicated above. The 30 year period of 1981 – 2010 would show an even greater warming and jump the hardiness zones another five degrees.

EXTREME DAILY TEMPERATURE FREQUENCIES

Another measure of the warming that has occurred in the 1981 – 2010 period is a comparison of the frequency of occurrence of new daily extreme high and low temperatures.

An examination shown in Table 44 reveals that for nearly all months the number of new extreme daily highs has exceeded new extreme daily lows. Assuming no trend in temperatures, equal numbers would be expected. For January only Poughkeepsie with its 82 year broken record has had more record lows. For February record highs greatly exceed record lows. For March and April there were more extreme highs than lows. In May and June only West Point showed an equal number of highs and lows. In July, Gloversville, Indian Lake and West Point had equal numbers of extreme highs and lows. In August only Indian Lake had a greater number of lows. Albany, Glens Falls, West Point and Poughkeepsie has an equal number. In September, Troy and Gloversville were out of step compared to the other stations. October for an unknown reason showed the greatest number, with seven stations showing more extreme lows than highs. November and December had all stations with more extreme highs than lows. Station relocations at Poughkeepsie, Glens Falls, Saratoga Springs and Troy may explain the greater number of record lows for certain months. Table 41 shows the hottest daily temperatures with month and year of occurrence.

FREQUENCY OF DAILY MINIMUM EXTREMES

The expected number of daily extremes for a station with 100 years of observations is an equal number 3-4 (each) of highs and lows per month assuming no temperature trend. For the 30 year period (1981-2010) this would be 3 new record highs and 3 new record lows. At seven long term stations only October shows a greater number of new record lows in the 1981-2010 decade. One possible explanation for this is that October is a transition month between summer and winter with a great variability of high and low temperatures. For other stations there have been only six other months with more record lows for the 1981 - 2010 period.

TABLE 44
NUMBER OF NEW DAILY RECORD MAXIMUM/ MINIMUM TEMPERATURES
FOR THE 1981 – 2010 PERIOD

	J	F	M	A	M	J	J	A	S	O	N	D
Albany 1874 -	9/4	16/3	8/4	10/4	8/7	3/2	2/5	1/1	2/1	0/5	8/3	13/3
Glens Falls C&AP 1893-1933 1943-	9/12	19/6	11/6	12/4	5/6	12/2	9/3	3/3	4/2	0/1	10/3	12/4
Gloversville 1893-1998 2001-06	11/10	13/2	11/6	12/3	6/3	4/2	0/0	4/3	1/1	1/5	7/3	9/3
Indian Lake 1900-	11/2	7/3	6/4	7/0	1/0	2/0	0/0	0/1	3/0	1/0	8/1	12/0
Middletown 1893-1908 1951-	22/10	19/5	17.9	17/5	9/3	7/4	14/3	12/3	13/3	5/2	15/1	19/3
Mohonk Lake 1895-	9/9	14/3	14/8	13/0	19/3	19/4	23/0	17/5	14/1	2/3	6/5	15/6
Poughkeepsie C&AP 1892-1901 1928-	4/5	7/2	10/6	9/4	7/3	1/1	5/3	4/4	8/5	2/4	7/5	12/6
Saratoga Spgs 1895-1906 1941-	17/13	21/4	12/3	18/3	16/5	14/4	15/4	5/2	13/2	3/6	14/2	15/4
Troy 1826-46 1913-35 1956-	9/7	15/3	12/8	11/3	3/5	5/1	4/3	7/6	2/3	3/1	11/8	14/4
West Point 1895-	7/6	13/1	13/5	8/3	4/4	4/5	3/3	2/2	7/4	1/7	7/6	12/2
Yorktown Hts/ Carmel	10/2	12/0	13/1	13/1	3/0	6/0	7/0	4/1	5/0	1/0	7/1	13/1

TABLE 45
DECADAL AVERAGES
In days of ice on period

	Mohonk Lake	Lake Mohansic	Lake Mahopac
1941 – 1950	108		
1951 – 1960	103		
1961 – 1970	113		
1971 – 1980	108	83	
1981 – 1990	99	79	77
1991 – 2000	89	67	59
2001 – 2010	98	73	54

HOT NIGHTS IN THE OLD TOWNS

FREQUENCY OF THE HOTTEST MINIMUM TEMPERATURES

Whenever the hottest summer temperatures occur, the question most often asked is whether a new record high has been set. For northern Westchester and Putnam counties, daily temperatures in the high 90s for the last one hundred and twenty years are usually the hottest of record for any particular day. There have also been nine days in that time period when maximum temperatures reached 100 or higher in the month of July and eight such days in August..

There is now a universal concern of global warming and many studies have shown a significant average warming over the last thirty years, plus a far greater frequency of new record high temperatures compared with record lows. However little or no attention has been given to the frequency and number of nights with minimum temperatures that were record breakers for heat. The following table shows the frequency in summer season months when the night time lows did not fall below 70 degrees.

The U.S. Cooperative Weather Service stations at Carmel and Yorktown Heights with the longest combined continuous record of daily observations dating back here to 1896 provide the data source for high and low temperatures (Table 46).

Two other Hudson Valley cooperative weather stations are listed showing the periods when 70 degree minimums have been equaled or exceeded.

In the years before home air conditioning, a night with low temperatures at 70 or above was an agonizing sleepless one. City apartment house dwellers in many cases slept on fire escapes hoping for even a slight breeze to relieve the heat.

The following Yorktown and Mohonk Lake record shows the number of days when the minimum temperature did not fall below 70 degrees. What is especially significant is the greater number of such events for each summer month in the past thirty years. Since 1978 every year but one has had at least one night with 70+ lows.

Of the twenty-six years in June at Yorktown with lows 70 and higher, thirteen came after 1980. Of the forty-eigtht years in July having such days, twenty-seven occurred after 1980. In August, of the forty-seven years recording such minimums, twenty came after 1980.

Whether through global warming, increased urbanization in our local region or most likely both, the pattern of greater night time heat in the last few decades for our region is quite clear. Table 46 shows the total number of such events for the summer season at Yorktown and West Point.

At the West Point station the record ending in 2012 has the greatest number of 70+ nightly minimums for the climatological summers of June, July and August for the most recent period.

At the Mohonk Lake station the same pattern of warm night minimum temperatures is shown for the 30 year period ending in 2010.

Successive summer months with such nighttime heat have rarely been experienced before. The warmest minimum temperatures of the last three decades are another indication of the regional warming in the Hudson Valley watershed.

TABLE 46
SUMMERS (J, J, A, S) OF DAYS WITH MINIMUMS OF 70 DEGREES AND HIGHER
1895 - 2012

Year	CARMEL/YORKTOWN	WEST POINT	MOHONK LAKE
1898	14	13	13
1901	18	14	10
1908	9	8e	5
1938	3	20	6
1949	6	14	16
1955	4	24	22
1983	7	7	11
1988	13	22	8
1995	10	15	5
1999	12	19	13
2002	14	18	17
2003	9	13	4
2005	27	28	18
2006	13	9	12
2008	10	12	8
2010	19	23	12
2011	9	13	5
2012	14	12	10

EXTREME PRECIPITATION

Along with the increase in temperature as shown above, has come an increase in the number and frequency of extremes of precipitation for all time periods since 1980. Greater warmth results in more and longer retention of atmospheric moisture giving this region greater totals of precipitation of all kinds.

When the extreme precipitation regime is examined for the 30 year, annual, monthly and daily periods, the pattern of extremes is similar to the pattern shown by temperature. Nearly all Hudson Valley watershed weather station rain gauges record their greatest amounts since 1981.

While thirty year divisional precipitation averages show an increase in the most recent period for two of the three divisions (Table 47), four of five of the oldest long term stations have had their wettest amounts in the most recent thirty years (Tables 48- 52). Eight of seventeen stations have had their wettest years since 2000 and all but five since 1983.

When noting the wettest events that have occurred since 1981 it is important to know the length of record. A station with a sixty year record can expected to have by chance alone five of its wettest ten events in the period from 1981 to 2011, while a station with a 100 year record can be expected to have three wettest events in the wettest ten.

For the greatest annual amounts as shown in Table 53 over half of the fifteen stations have their wettest years since 1981. Only Albany City and Airport with its 185 year record shows only two years (2005) and 2011within its ten wettest since 1981. The stations with half or more of their wettest years could not have had these occur by chance.

Within each of the months most stations had at least half of their wettest within the most recent thirty one year period of record (Tables 54 – 68).

An examination of the ten wettest days for their periods of record shows that even those stations with over 100 years of daily records have had at least one day falling within the 1981 – 2011 period in any given month, thus exceeding what would be expected by chance.

An examination of rank order of greatest daily precipitation by month at eleven stations shows that six have had their wettest days since 1981 (Tables 69 - 79).

Table 80 is a summary of the wettest days for each month for seventeen stations.

TABLE 47
PRECIPITATION
DIVISION - 30 YEAR AVERAGES
(Inches)

Hudson Valley

	Jan	Feb	Mar	Apr	May	Jun	Jul	Aug	Sep	Oct	Nov	Dec	Annual
1971 – 2000													
	3.34	2.55	3.56	3.86	4.40	4.09	4.16	4.07	4.02	3.61	3.75	3.28	44.60
1981 - 2010													
	3.11	2.61	3.54	3.74	4.07	4.35	4.36	4.12	4.03	4.28	3.72	3.60	45.53

Mohawk Valley

	Jan	Feb	Mar	Apr	May	Jun	Jul	Aug	Sep	Oct	Nov	Dec	Annual
1971 – 2000													
	3.42	2.69	3.50	3.73	3.92	4.28	4.24	4.05	4.65	3.65	4.17	3.64	45.94
1981 - 2010													
	3.15	2.61	3.32	3.74	4.05	4.37	4.19	4.28	4.11	4.17	3.96	3.64	45.60

Northern Plateau

	Jan	Feb	Mar	Apr	May	Jun	Jul	Aug	Sep	Oct	Nov	Dec	Annual
1971 – 2000													
	3.60	2.72	3.63	3.41	3.80	4.00	4.15	4.28	4.61	3.88	4.25	3.77	45.80
1981 - 2010													
	3.35	2.75	3.09	3.37	3.88	4.17	4.23	4.23	4.39	4.57	4.16	3.84	46.12

TABLE 48
GREATEST ANNUAL PRECIPITATION
YORKTOWN 1860 - 2011
1971 - 2000 Normal = 51.00 Inches
1981 - 2010 Normal = 51.23 Inches

Years Greater than 55.00 Inches		Years Greater then 60.00 Inches
1868	1972	1907
1882	1975	1927
1888	1979	1937
1898	1983	1955
1901	1984	1972
1907	1990	1979
1919	1994	1983
1920	1996	1994
1927	2003	1996
1937	2005	2005
1938	2006	2011
1942	2008	
1945	2011	
1952		
1955		

TABLE 49
WEST POINT
Greatest Annual Precipitation 1840 -2011
1971 - 2000 Normal = 50.30 Inches
1981 - 2010 Normal = 50.62 Inches

Years Greater than 55.00 inches		Years greater than 60.00 Inches
1850	1983	1853
1853	1990	1952
1889	1994	1979
1890	1996	1983
1913	2003	1996
1951	2010	2003
1952	2011	2011
1955		
1960		
1972		
1975		
1979		

TABLE 50
GREATEST ANNUAL PRECIPITATION
ALBANY CITY & AIRPORT 1826 - 2011
1971 – 2000 Normal = 38.60 Inches
1981 - 2010 Normal = 39.35 Inches

Years equal or greater than 45 Inches			Years equal or greater than 50 Inches
1836	1972	2000	1850
1842	1973	2003	1870
1843	1975	2005	1871
1847	1977	2006	1975
1850	1979	2008	1977
1870	1983	2011	1979
1871	1989		1983
1890	1996		2011
1951			
1960			

TABLE 51
GREATEST ANNUAL PRECIPITATION
MOHONK LAKE 1899 - 2011
1971 – 2000 Normal = 51.00 Inches
1981 - 2010 Normal = 51.86 Inches

Years equal or greater than 55 Inches		Years equal or greater than 60 Inches
1901	1983	1902
1902	1989	1945
1903	1990	1983
1906	1996	1990
1915	2000	1996
1934	2003	2003
1945	2005	2005
1952	2006	2008
1955	2008	2009
1972	2009	2011
1973	2011	
1975		
1979		

TABLE 52
GREATEST ANNUAL PRECIPITATION
INDIAN LAKE 1899 - 2011
1971 - 2000 Normal = 41.75 Inches
1981 - 2010 Normal = 40.85 Inches

Years Equal or greater than 45 Inches		Years Equal or greater than 50 Inches
1902	1972	1932
1929	1976	1945
1932	1977	1951
1942	1979	1954
1945	1983	1972
1946	1990	1976
1951	2000	2006
1954	2005	2008
	2006	2011
	2008	
	2010	
	2011	

TABLE 53
WETTEST YEARS
(Inches)

	1	2	3	4	5	6	7	8	9	10	
Albany City & Airport 1826 -											
	55.18	54.49	54.18	53.67	52.89	51.82	50.08	49.80	49.02	48.91	
	1977	1871	1870	2011	1850	1979	1975	1827	1973	1843	
Alcove Dam 1948 -											
	63.69	60.07	55.18	54.66e	51.82	51.13	50.08	49.02	48.58	47.15	46.70
	2011	1983	1977	1996	1977	2006	1975	1973	2000	2008	1955
Glens Falls AP 1944 -											
	59.95	57.00	56.68	55.86	54.40	54.18	52.82	52.79	50.76	50.34	
	1990	2008	2006	1975	1983	1972	2011	1996	1977	1973	
Middletown 1893-1907 1952 –2008											
	57.09	56.82	55.89	54.36	54.04	53.70	52.14	51.98	51.24	51.21	
	1952	1903	1901	1996	1975	1983	1990	2003	2006	1955	
Millbrook 1942 –											
	64.59	56.14	54.37	54.31	54.02	53.80	52.41	51.16	49.55	49.41	
	2011	1996	2003	1990	2005	1972	1945	1955	1973	1977	
Mohonk Lake 1891-											
	68.80	64.30	63.26	62.79	62.53	62.41	62.29	60.75	60.43	60.12	
	1996	2011	2005	1990	2003	1983	1945	1902	1901	2008	
Peekskill 1877-97 1926 –											
	71.76	68.48	62.74	62.57	62.45	61.80	61.43	60.40	60.19	59.79	
	1990	1983	2011	1996	1888	1975	2003	1945	1889	1952	
Poughkeepsie 1830-49 1928-											
	61.17	60.91	60.25	60.0e	59.25	55.44	49.84	49.73	49.16	48.49	48.12
	2003	1996	1945	2005	2011	1983	2006	1831	1990	1952	2008
Saratoga Springs 1941 –											
	63.13	57.37	55.53	53.23	52.77	52.69	52.42	51.93	50.14	50.0e	
	2011	2008	1983	1972	2006	2005	1990	1996	2003	1945	
Troy 1913-35 1956 – (missing 2011)											
	52.31	48.71	48.70	48.31	47.70	47.69	46.8e	46.56	46.14	45.70	
	2008	2003	1975	2004	2006	1996	2005	1983	1977	2007	

	1	2	3	4	5	6	7	8	9	10

Westchester
Airport 1949-

| 74.15 | 69.83 | 63.67 | 61.83 | 61.29 | 61.04 | 60.36 | 60.24 | 69.34 | 58.96 | |
| 1983 | 2011 | 1972 | 1996 | 1984 | 1951 | 1975 | 1852 | 1955 | 1989 | |

West Point 1840 –

| 78.37 | 71.39 | 69.49 | 67.72 | 65.60 | 65.01 | 63.56 | 62.44 | 60.0e | 59.94 | |
| 2011 | 1996 | 1983 | 2005 | 2003 | 1952 | 1853 | 1979 | 1901,02 | 1850 | |

Yorktown 1860-

| 74.74 | 66.03 | 64.15 | 64.03 | 63.95 | 63.80 | 62.62 | 62.29 | 61.91 | 61.74 | |
| 2011 | 1996 | 1927 | 1983 | 1955 | 2005 | 1937 | 1952 | 1979 | 1907 | |

Gloversville 1892 -2007

| 60.23 | 58.08 | 57.66 | 56.77 | 55.10 | 52.89 | 51.96 | 51.85 | 49.10 | 49.02 | |
| 1972 | 2003 | 1977 | 2006 | 1945 | 1975 | 1990 | 1996 | 1955 | 1983 | |

Little Falls 1897 –2010

| 55.08 | 53.37 | 51.69 | 50.52 | 48.43 | 48.0e | 47.41 | 46.83 | 45.87 | 45.00e | |
| 1950 | 1990 | 1945 | 1972 | 1977 | 2003 | 1996 | 1975 | 2006 | 2008 | |

Utica 1911 –2009

| 61.62 | 54.15 | 52.09 | 51.04 | 48.68 | 48.53 | 48.30 | 48.26 | 47.42 | 47.25 | |
| 1972 | 1990 | 1975 | 1971 | 1927 | 1986 | 1929 | 1984 | 1945 | 1996 | |

Indian Lake 1899 –

| 56.47 | 55.30 | 53.45 | 52.45 | 52.15 | 51.15 | 51.03 | 50.59 | 48.91 | 45.96 | 45.79 |
| 2011 | 1945 | 1976 | 2008 | 1951 | 1972 | 1954 | 2006 | 1947 | 2005 | 1990 |

TABLE 54
WESTCHESTER AIRPORT
Wettest Months / Years 1949 - 2012

	1	2	3	4	5	6	7	8	9	10	
January	11.17	8.17	8.05	7.60	6.08	5.97	5.94	5.92	5.88	5.51	
	1979	1978	1958	1983	1982	1999	1976	1953	1952	1986	
February	7.25	7.00e	6.10	5.75	5.52	5.32	5.25	5.21	5.01	4.99	
	1981	2008	1961	1962	1950	1971	1951	1972	1960	1956	
March	11.44	11.34	11.10	10.00e	8.69	7.75	7.67	7.40	6.77	6.04	5.84
	1953	1980	1983	2010	1951	1977	1967	2011	1984	1999	1994
April	13.98	12.11	9.43	8.37	8.08	8.07	7.86	7.64	7.46	7.10	6.72
	2007	1983	1952	1984	1958	1973	1961	1996	1980	2011	1982
May	13.29	11.22	8.36	8.18	7.44	7.08	6.72	6.46	5.64	5.56	
	1989	1984	1952	1978	1972	1968	2011	1990	2012	1998	
June	14.29	8.12	6.85	6.40	6.39	6.21	6.19	6.13	5.94	5.91	
	1972	2003	1982	2009	1976	1992	2006	2001	2011	1998	
July	8.83	7.38	7.31	7.26	6.61	6.11	6.07	5.67	5.50	5.42	
	1984	1960	1969	1975	2000	1996	1997	1956	1994	1973	
August	13.97	13.13	10.46	8.45	8.21	8.14	7.57	7.43	7.35	6.96	
	2011	1955	1990	1954	2011	1952	1991	1962	1971	1976	
September	14.08	13.79	12.84	11.07	10.29	8.42	8.12	7.74	7.71	7.04	6.69
	2004	2011	1975	1974	1999	1996	1954	1977	1971	1966	2003
October	17.21	13.85	10.20	9.81	8.27	8.12	7.15	6.99	6.07	5.83	
	2005	1955	1983	1989	1995	1996	1990	1959	1956	1958	
November	9.24	8.95	8.89	8.45	7.34	7.25	7.03	6.87	6.68	6.47	6.09
	1972	1985	1977	1963	1954	1951	1950	1955	1988	1992	2007
December	10.09	9.38	8.23	7.51	7.34	6.86	6.63	6.41	6.03	6.02	
	1983	1973	1967	1996	1957	1990	1969	1986	1992	1959	
Annual	74.15	69.83	63.67	61.83	61.29	61.04	60.36	60.24	59.34	58.96	58.57
	1983	2011	1972	1996	1984	1951	1975	1952	1955	1989	1977

TABLE 55
YORKTOWN 1860 – 2012
WETTEST MONTHS

	1	2	3	4	5	6	7	8	9	10
January	11.29 1979	8.11 1891	7.74 1978	7.45 1996	7.40 1999	7.37 1905	7.20 1910	6.47 1949	6.50 1937	6.44 1887	
February	8.09 2008	7.70 1900	7.60 1981	7.11 1883	7.08 1887	6.70 1881	6.57 1909	6.52 1886	6.37 1870	6.25 1915	
March	10.77 1980	9.97 1953	9.32 2010	8.98 1877	8.13 1983	7.97 1912	7.95 2011	7.75 1977	7.71 1896	7.39 1942	
April	10.23 1983	9.81 2007	8.97 1901	8.62 1987	7.93 1924	7.92 1952	7.68 1980	7.51 1909	7.45 1984	7.38 1918	7.27 2011
May	11.04 1989	10.84 1984	9.93 1978	9.54 1947	8.88 1993	8.79 1868	8.60 1898	8.48 1952	7.98 1948	7.94 1946	
June	13.62 1972	10.07 1994	9.45 2009	9.22 1903	9.09 1928	8.61 2001	8.60 2006	7.64 1989	7.54 1938	7.44 2003	
July	11.46 1945	11.10 1960	10.99 1897	10.72 1877	10.60 1984	10.37 1996	10.27 1863	10.24 1950	10.13 1938	9.74 1946	
August	15.59 2011	14.45 1955	13.05 1937	11.13 1933	10.41 1952	10.33 1875	10.12 1990	9.72 1927	9.04 1971	8.96 1954	
September	18.00 1999	16.26 1882	13.17 2004	11.85 1907	10.98 1938	10.77 1975	10.08 1934	9.42 1954	9.33 1868	9.00 1974	
October	17.24 2005	15.21 1955	10.69 1877	9.16 1995	9.00 1913	8.93 1869	8.70 1943	8.16 1959	7.86 1989	7.26 1996	
November	9.02 1988	8.90 1947	8.59 1977	8.44 1889	8.21 1947	8.16 1972	7.98 1963	7.91 1932	7.54 1877	7.43 1954	
December	9.18 1973	8.59 1983	7.74 1901	7.62 1878	7.59 1996	7.54 1881	7.20 2008	7.17 1957	6.91 1967	6.84 1972	
Annual	74.74 2011	66.03 1996	64.15 1927	64.03 1983	63.95 1955	63.80 2005	62.62 1937	62.29 1952	61.91 1979	61.74 1907	

TABLE 56
PEEKSKILL 1926 - 2012
WETTEST MONTHS / YEARS

	1	2	3	4	5	6	7	8	9	10
January	11.34 1979	7.80 1999	7.50 1978	7.07 1936	7.02 1996	6.25 1937	6.10 1949	5.40 1984	5.28 2006	5.21 1976
February	8.66 1981	5.79 1971	5.55 2008	5.49 1976	5.25 1962	5.01 1951	4.95 1970	4.74 1969	4.57 1973	4.44 1950
March	9.42 1953	8.21 1980	7.81 1983	7.75 1951	7.71 2010	6.55 2011	6.45 1940	6.34 1993	6.11 1942	6.04 2001
April	11.19 1983	8.75 1952	7.76 1929	7.48 1980	7.10 1987	7.04 1984	6.60 1973	6.28 1996	6.24 2011	6.09 1933
May	11.70 1984	11.53 1989	9.85 1978	8.44 1998	8.00 1947	7.28 1946	7.21 1990	6.78 1968	6.64 1949	6.39 1979
June	9.72 2001	9.23 1928	9.04 1972	7.87 2009	7.79 1995	7.42 1989	7.25 1982	6.46 1968	6.36 2003	6.32 1931
July	13.85 1945	10.19 1960	9.75 1938	8.50 1996	8.18 1928	8.05 1975	7.88 1934	7.87 1946	7.63 1950	7.18 1990
August	16.71 1990	15.16 2011	12.97 1955	11.34 1933	10.65 1937	10.34 1926	8.74 1954	8.50 2004	8.45 1971	8.39 1952
September	15.02 1999	11.89 2004	11.52 1934	11.41 1975	10.62 1938	8.80 1977	8.25 1954	7.86 2003	7.42 1991	7.28 1996
October	15.04 2005	12.84 1995	12.29 1955	9.21 1989	8.12 1943	7.90 1927	7.19 1959	7.08 1976	6.79 1983	6.62 1996
November	9.65 1977	8.57 1972	8.41 1988	8.21 1932	7.47 1947	7.37 1954	7.06 1951	6.85 1963	6.81 1927	6.78 1992
December	10.44 1973	8.74 1983	8.10 1957	7.50 2003	7.13 1996	6.54 1953	6.43 1967	6.41 1936	6.22 1948	6.01 1969
Annual	71.76 1990	68.48 1983	62.74 2011	62.57 1996	62.45 1888	61.80 1975	61.42 2003	60.40 1945	60.19 1889	58.40 1952	58.15 1955

TABLE 57
WEST POINT 1840 – 2012
WETTEST MONTHS

	1	2	3	4	5	6	7	8	9	10	
January	11.64	11.07	9.95	9.25	8.20	7.81	6.93	6.64	6.23	6.17	
	1996	1979	1841	1978	1891	1999	1892	1953	2005	1958	
February	8.58	7.49	7.29	7.0e	6.29	6.22	6.15	5.88	5.60	5.45	
	1981	2008	1893	1900	2010	1847	1870	1876	1887	1853	
March	12.02	10.32	9.16	8.85	8.71	8.54	8.02	7.83	7.45	7.38	
	1896	1913	2011	1953	1876	1983	1999	1951	1899	1980	
April	10.53	10.63	9.47	9.45	9.0e	7.68	7.43	7.25	7.24	6.99	
	1854	1983	1952	1910	1901	1980	1973	1929	1851	2011	
May	11.66	11.18	11.09	10.50	9.68	8.26	8.24	8.01	7.99	7.75	
	1868	1984	1989	1867	1978	1850	1893	1972	1853	1865	
June	9.76	9.56	9.2e	8.28	7.46	7.37	7.36	7.24	7.14	7.11	
	2009	1972	1903	1989	1982	1848	1871	1994	2000	2003	
July	13.05	10.21	10.48	10.0e	9.92	9.64	9.00	8.65	8.55	8.40	
	1897	1945	1853	1863	1889	1960	1996	1984	1871	1884	
August	17.64	11.96	11.75	11.74	11.70	11.40	11.33	10.68	9.92	9.08	
	2011	1856	1868	1955	1867	1933	1843	1990	1842	1898	
September	15.05	13.50	12.30	12.04	11.22	10.74	10.68	9.74	9.45	9.33	
	2003	1882	1999	2004	1868	1938	1907	1975	2011	1924	
October	21.10	12.75	12.21	10.97	10.25	8.76	8.66	8.54	8.49	8.33	7.86
	2005	1955	1995	1959	1855	2010	1913	1877	1911	1840	1989
November	10.02	9.01	8.53	8.52	8.30	8.08	7.97	7.93	7.42	7.28	
	1846	1972	1977	1988	1932	1889	1947	1954	1979	1969	
December	11.47	10.17	9.51	8.12	8.0e	7.94	7.81	7.67	7.03	6.97	
	1983	1996	1957	1973	1901	1948	2008	2009	1990	1878	
Annual	78.37	71.39	69.49	67.72	65.60	65.01	63.56	62.44	60.0e	59.94	
	2011	1996	1983	2005	2003	1952	1853	1979	1901,02	1850	

TABLE 58
MIDDLETOWN 1893 – 1907, 1952 – 2008
2NW & 2 (1972 – 2008 WettestMonths / Years (Inches)

	1	2	3	4	5	6	7	8	9	10
January	6.33 1953	6.28 1979	6.17 1978	6.00 2005	5.72 1996	5.16 1990	4.79 1976	4.77 1975	4.75 1999	4.50 2006
February	7.05 2008	6.60 1893	6.51 1896	5.74 1981	5.69 1900	4.47 1962	4.06 1971	3.27 1982	2.94 2003	2.66 1990
March	8.09 1896	7.15 2011	6.54 1977	6.24 2008	6.04 1983	5.86 1901	5.13 1984	5.00e 1993	4.98 1994	4.53 1999
April	9.62 1983	9.11 1901	8.33 1952	6.97 2007	6.49 1996	6.45 1973	6.18 2011	6.05 1958	5.99 2005	5.77 1953
May	10.04 1984	9.94 1989	7.10 1968	6.90 1901	6.76 1978	6.70 1953	6.56 2002	6.51 1990	6.21 1998	5.95 1983
June	12.00 1903	11.61 2009	10.38 1998	10.23 1972	9.78 2006	7.70 1973	6.64 2003	6.57 1952	6.48 1989	6.17 1992
July	10.61 1988	9.14 1897	8.87 1996	8.06 1963	7.68 1959	7.39 1975	6.80 1992	6.61 2000	6.27 1984	5.91 1987
August	12.89 1955	9.13 1990	9.04 1903	8.50 1901	8.24 2004	6.96 2009	6.89 1971	6.23 1976	5.78 1960	5.68 1893
Septembe	8.42 1907	8.35 2004	8.24 1999	8.22 1977	7.91 2003	7.89 1899	7.35 1996	7.32 1979	7.21 1952	7.08 1960
October	14.00e 2005	11.52 1955	11.39 1959	9.05 1903	8.49 1995	7.05 2006	7.04 2010	6.55 1976	6.01 2003	5.94 1958
November	7.73 1972	6.83 1954	5.92 1988	5.72 1977	5.50 1992	5.05 1963	4.50 1969	4.14 1994	4.03 2005	3.81 1997
December	8.79 1973	6.74 1901	6.32 1990	6.13 2008	5.99 1957	5.97 1996	5.73 2007	5.51 1983	5.49 1953	5.06 1969
Annual	57.09 1952	56.82 1903	55.89 1901	54.36 1996	54.04 1975	53.70 1983	52.14 1990	51.98 2003	51.24 2006	51.21 1955

TABLE 60
MOHONK LAKE 1891 - 2012
Wettest Months / Years

	1	2	3	4	5	6	7	8	9	10	
January	11.24	9.63	7.82	7.28	7.11	6.99	6.64	6.52	6.27	5.37	
	1979	1891	1978	1910	2005	1999	1953	1996	1958	1994	
February	10.09	8.93	7.06	6.69	6.38	6.33	6.28	6.00	5.74	5.70	
	2008	1893	1910	1981	2010	1926	1909	1900	1971	1920	
March	11.07	8.95	8.77	8.76	8.48	8.34	8.25	7.90	7.67	7.06	
	1896	1914	2008	1977	1983	1936	2010	1951	1953	2011	
April	11.32	10.52	9.39	8.90	8.41	8.33	8.03	7.69	7.18	6.94	
	1901	1983	1952	1929	1987	2007	1910	1996	1973	2011	
May	14.12	10.92	9.55	8.75	8.44	8.42	8.34	8.03	7.60	7.46	
	1989	1984	1908	1901	1968	1978	1953	1893	1942	1998	
June	13.66	12.44	9.45	9.30	9.05	8.73	8.72	8.19	5.97	5.14	
	2009	1903	1938	1972	1998	2006	1928	1973	2003	2001	
July	14.71	11.18	10.00	9.75	9.15	9.14	8.19	7.68	7.08	6.00	
	1897	1945	1996	1902	1891	1975	1995	2009	1960	1984	
August	16.76	14.11	10.80	10.70	10.19	9.72	9.53	8.73	8.70	8.19	8.02
	2011	1955	1903	1933	1971	1928	1901	2004	1990	1898	2009
September	12.08	11.59	11.17	10.01	9.92	9.65	9.01	8.90	8.68	8.58	8.56
	2011	1938	1987	1934	1996	1999	1907	1899	2003	1960	2004
October	14.09	11.09	10.51	9.43	8.90	8.21	8.18	7.72	7.02	6.95	
	1955	1995	1911	1959	1903	2005	1976	2006	1989	1996	
November	9.32	8.91	8.90	8.26	7.38	6.44	6.39	6.16	6.14	5.83	
	1972	1932	1891	1927	1977	1954	1997	1969	1988	1947	
December	9.23	9.22	8.92	8.90	8.21	8.05	7.88	7.07	6.91	6.88	
	1948	1973	1996	1915	2008	1990	1957	1983	2007	1901	
Annual	79.31	68.80	63.26	62.79	62.53	62.41	62.29	60.75	60.43	60.12	
	2011	1996	2005	1990	2003	1983	1945	1902	1901	2008	

TABLE 61
POUGHKEEPSIE City and Airport
Wettest Months / Years

	1	2	3	4	5	6	7	8	9	10
January	8.30 1979	8.0e 1978	7.41 1996	6.65 1891	6.42 1937	6.16 1999	6.14 1836	5.35 1892	5.23 1953	4.88 1832
February	7.51 2008	6.78 1893	6.25 1900	5.42 1981	4.58 1951	3.88 1971	3.78 1939	3.65 2011	3.53 1972	3.50 1945
March	7.39 1983	7.11 1953	6.25 2008	6.20 1896	5.97 1980	5.55 1899	5.51 1890	5.36 1999	5.17 2011	5.10 1951
April	8.51 1983	8.09 2007	7.57 1952	7.48 1929	6.50 1973	6.47 1987	6.22 1996	6.03 1933	6.01 1958	4.99 1957
May	11.81 1989	11.49 1984	7.74 1972	7.49 1945	7.05 1893	7.00 1978	6.64 1968	6.36 1849	6.31 2009	5.33 1952
June	9.22 2009	8.39 1982	7.99 1972	7.83 1928	7.66 1994	7.33 2006	7.29 1998	7.21 1989	7.02 2000	6.72 1968
July	13.63 1975	12.92 1996	10.77 1945	9.37 1897	8.68 1969	8.45 1844	7.19 1928	7.08 1938	6.38 2009	6.01 1956
August	13.23 2011	12.71 1955	10.92 1971	9.05 1928	8.85 1898	8.71 1933	8.67 1931	8.26 1932	7.91 1990	7.27 1976
September	12.59 2003	9.89 1934	9.85 1938	8.85 2004	7.50 2011	7.43 1930	7.19 1975	6.94 1999	6.85 1960	6.67 1966
October	18.17 2005	10.40 1955	9.98 1995	8.64 1833	8.20 1843	6.81 1959	6.78 1976	6.54 1989	5.99 1894	5.60 1996
November	8.31 1932	8.11 1972	7.2e 1977	7.10 1988	7.00 1930	6.67 1846	6.41 1969	6.01 1997	6.77 1954	5.96 1892
December	8.65 1973	7.40 1948	7.28 2003	7.09 1996	6.96 2008	6.60 1983	6.46 1847	5.85 1830	5.71 1957	5.68 1969
										5.00 1990
Annual	61.17 2003	60.91 1996	60.29 2011	60.25 1945	60.0e 2005	55.44 1983	49.84 2006	49.73 1831	49.16 1990	48.49 1952
										48.12 2008

TABLE 62
ALBANY CITY 1826 - 1940 · ALBANY AIRPORT 1941 - 2012
WETTEST MONTHS / YEARS (Inches)

	1	2	3	4	5	6	7	8	9	10	
January	7.62	6.69	6.66	5.85	5.40	5.17	5.09	5.08	4.96	4.84	
	1836	1891	1978	1979	1827	1987	1870	1923	1962	1835	
February	6.16	5.06	5.04	4.91	4.63	4.58	4.57	4.52	4.40	4.37	
	1981	1893	2008	1870	1926	1971	1851	1891	1896	1909	
March	9.14	7.85	6.64	6.63	6.56	6.25	6.21	5.74	5.50	5.24	
	1980	1977	1983	1843	1871	1936	2008	1953	2001	1888	
April	9.68	7.44	6.33	6.14	6.08	5.88	5.38	4.96	4.87	4.77	
	1983	1987	1857	1854	1973	1929	1914	1984	1924	1829	
May	8.96	8.47	8.20	7.83	7.30	7.05	6.71	6.61	6.45	6.33	
	1953	1833	1984	1989	1848	1973	1838	1837	1853	1868	
June	10.15	9.03	8.56	8.52	8.36	8.22	8.10	7.58	7.40	7.33	
	1983	1928	1972	1862	1973	1838	1989	1830	1855	1870	
July	9.91	9.33	9.27	9.18	8.74	8.57	8.41	8.36	8.08	7.51	
	2009	1945	1850	1871	2006	1848	1960	1975	1981	1858	
August	10.97	10.41	10.38	8.33	8.30	8.26	7.92	7.51	7.37	7.36	7.34
	1856	2011	1871	1885	1870	1955	1893	1832	1927	1892	2004
September	11.06	10.41	9.32	8.50	8.08	8.01	7.89	7.46	7.31	6.62	
	1999	1890	1938	1977	1828	1987	1960	1955	1868	2011	
October	13.87	9.18	9.00	8.83	8.27	8.03	7.50	7.19	7.10	6.97	
	1869	1855	2005	1955	1849	1995	1833	1989	2010	1877	
November	7.98	7.83	7.66	7.29	6.21	6.20	6.11	6.10	5.91	5.72	
	1972	1927	1977	1830	1988	1921	1932	1886	1997	1969	
December	8.35	6.94	6.51	6.50	6.05	5.93	5.92	5.83	5.61	5.48	
	1973	1915	1969	1983	1948	1887	1901	1878	1902	2003	
Annual	60.07	55.18	54.49	54.18	53.68	52.89	51.82	50.08	49.80	49.02	
	1983	1977	1871	1870	2011	1850	1979	1975	1827	1973	

TABLE 63
ALCOVE DAM
1948 – 2012
Wettest Months / Years

	1	2	3	4	5	6	7	8	9	10	
January	6.94 1996	6.46 1978	5.85 1979	5.29 2006	5.08 1995	4.81 1954	4.76 1999	4.37 2005	3.66 1976	3.59 1990	
February	6.16 1981	5.58 2010	5.37 2008	4.58 1971	4.12 1950	4.11 1949	3.98 1990	3.70 1972	3.36 1977	3.29 1962	
March	7.85 1977	6.91 2001	6.69 2008	6.64 1983	6.59 1953	5.80 1999	5.65 2011	4.86 1993	4.79 1960	4.73 1984	3.64 1994
April	9.68 1983	6.25 2007	6.18 1996	6.08 1973	5.88 2011	5.33 1952	4.96 1984	4.79 1980	4.40 1979	4.06 2004	
May	8.20 1984	7.83 1989	7.05 1973	6.62 1968	6.58 1998	6.53 1953	6.15 1954	6.06 2011	5.94 1990	5.42 1983	
June	10.15 1982	9.19 2006	8.56 1972	8.36 1973	8.23 1998	8.10 1989	5.26 2010	5.05 2011	5.00e 2003	4.95 1975	4.85 1952 / 4.81 2001
July	8.36 1975	7.39 1979	6.95 2009	6.18 2010	6.12 1988	6.00e 1996	5.83 2004	5.67 1987	4.98 2000	4.59 1954	4.17 1984
August	12.11 2011	9.41 1955	8.50e 1990	7.06 1971	6.36 2004	5.07 2009	4.95 2010	4.93 2000	4.62 1994	4.59 1973	
September	11.23 1999	9.37 1960	8.50 1977	8.16 2011	6.01 1979	5.53 2003	5.38 1996	5.25 1952	4.98 1971	4.68 1951	4.67 2004
October	14.54 1955	11.80 2005	10.47 2010	7.00e 1995	6.67 1975	6.18 1979	6.01 2006	5.74 1959	5.14 2007	5.04 1958	5.00e 2003
November	7.98 1972	7.66 1977	6.21 1988	6.10 1997	5.89 1969	5.30 1959	5.21 1954	5.04 1959	4.83 1951	4.54 2005	
December	8.35 1973	6.37 2008	6.29 1969	6.02 2003	5.75 1957	5.36 1996	5.35 1952	5.05 2007	4.75 1949	4.67 1948	
Annual	63.69 2011	60.07 1983	55.18 1977	54.66e 1996	51.82 1977	51.13 2006	50.08 1975	49.02 1973	48.58 2000	48.52 2010	46.70 1955

TABLE 64
TROY Lock & Dam 1913 - 1935   1956 - 2012
Wettest Months / Years

	1	2	3	4	5	6	7	8	9	10
January	6.77 1979	5.01 1978	4.99 2005	4.96 2006	4.78 1923	4.50 1996	3.70 1958	3.55 1976	3.13 2002	2.90 1994
February	5.28 1981	5.12 2008	3.86 1926	3.53 1984	3.46 1975	2.90 1994	2.85 1971	2.70 1976	2.55 1977	2.49 1962
March	8.33 2001	6.46 2008	6.14 1977	4.70 1993	4.45 1994	4.22 1983	4.16 2007	4.12 2005	3.65 1972	3.55 2002
April	7.23 1983	6.25 1929	5.63 1996	5.24 1993	5.18 2007	4.67 1933	4.48 2006	4.34 1962	4.30e 1987	4.27 1973
May	8.86 1984	7.05 2011	6.23 1996	6.20 2004	6.16 1983	6.00 1973	5.95 1989	5.58 1990	5.08 2000	5.02 2002
June	10.16 2006	7.85 1998	7.10 2002	6.97 1928	6.96 2008	6.91 1989	6.74 1972	6.52 1982	6.45 2000	6.08 1986
July	9.36 1984	8.67 2004	8.44 2009	8.12 1960	7.31 1996	7.27 1931	6.68 1975	6.41 2008	5.00 1976	4.96 1969
August	10.00e 2011	8.96 2004	8.01 1986	7.62 1994	7.57 1977	7.24 1928	7.14 1933	7.01 1971	6.22 1976	6.19 1990
September	8.63 1960	8.00e 2011	7.78 1933	6.38 1987	6.14 1977	5.93 2004	5.78 1999	5.64 1966	5.19 1979	5.17 1934
October	10.43 2010	9.14 2005	7.63 1995	7.46 2007	7.22 1987	5.86 1975	5.48 2008	5.20 1976	4.91 1927	4.67 2002
November	7.99 1927	6.89 1972	6.29 1959,69	5.41 1932	5.04 1983	4.99 1988	4.81 1968	4.77 1997	4.19 1995	4.00 1977
December	6.69 1973	5.82 2003	5.41 2008	5.14 1983	4.93 1957	4.89 1969	4.18 1977	3.72 1972	3.69 2007	3.64 1996
Annual	52.75e 2011	52.31 2008	48.71 2003	48.70 1975	48.31 2004	47.70 2006	47.69 1996	46.80e 2005	46.56 1983	46.14 1977

Note: April column 11: 4.22 1984; August column 11: 5.92 2009; Annual column 11: 45.70 2007

TABLE 65
GLENS FALLS 1896 – 1933 1944 – 2012 (Farm)
Wettest Months

	1	2	3	4	5	6	7	8	9	10	
January	7.49	6.37	6.27	6.26	6.25	5.10	4.96	4.80	4.76	4.72	
	1999	1978	2006	1979	1923	1996	1953	1910	1958	1994	
February	8.30	6.66	6.35	6.16	5.50	5.48	5.06	4.38	3.74	3.70	
	2008	1981	1909	1915	1893	1971	1910	1900	1926	1950	
March	8.29	7.31	6.52	6.38	6.20	6.13	5.97	5.72	5.42	5.22	
	1913	1977	1896	2011	1953	2001	2008	1899	1994	1980	
April	6.64	6.50	6.40	6.10	5.86	5.67	5.27	5.02	4.83	4.77	
	1924	1983	1996	1929	1914	2003	1990	1973	1984	2011	
May	9.66	8.92	8.46	7.31	7.16	6.63	6.00	5.92	5.78	5.59	
	1983	1984	1990	1989	1973	1947	1953	1972	1945	1910	
June	9.86	9.56	8.97	8.62	7.75	6.21	6.08	5.98	5.55	5.41	
	1944	1998	1903	2003	1972	1989	1994	1975	1928	2011	
July	8.82	8.31	8.30	7.94	7.48	6.78	6.45	6.30	6.14	5.89	5.81
	1897	1994	1996	1915	1960	2006	1902	2008	1944	1945	2009
August	11.00e	8.27	7.03	7.02	7.00	6.45	6.19	5.30	5.25	5.06	
	2011	1990	1955	1983	1898	1971	1933	2004	2010	1903	
September	8.60	8.13	8.04	7.61	7.59	7.10	7.02	6.94	6.61	6.59	
	1999	1987	1975	1907	2003	1899	1981	1945	2011	1977	
October	10.60	8.87	8.66	7.27	7.07	6.97	6.69	6.60	6.46	6.20	
	1955	2005	1995	1977	1976	2009	1911	2006	1959	2010	
November	8.08	7.51	7.11	6.90	6.01	5.89	5.77	5.69	5.61	5.26	
	1972	1983	1932	1927	1988	1921	1974	1969	1954	2005	
December	8.67	8.43	7.97	6.82	6.80	6.08	5.96	5.83	5.82	5.69	
	1973	2008	1948	1996	1983	1998	1927	1902	1901	1969	
Annual	59.95	59.80	57.00	56.68	55.86	54.40	54.18	52.79	50.76	50.34	
	1990	2011	2008	2006	1975	1983	1972	1996	1977	1973	

TABLE 66
SARATOGA SPRINGS (1941 – 2012)
Wettest Months / Years

	1	2	3	4	5	6	7	8	9	10	
January	6.64	6.63	6.21	6.12	5.78	5.20	4.49	4.34	4.00e	2.53	
	1998	1979	1950	1999	1978	1996	1976	1994	1953	1962	
February	6.25	5.78	4.74	4.49	3.87	3.85	3.60	3.24	3.19	3.00e	
	2008	1971	1981	1950	1951	1972	1945	1986	1976	1954	
March	6.36	6.22	6.05	5.81	5.65	5.10e	5.04	5.00e	4.43	4.33	
	2001	2008	1977	1997	1980	1955	1994	1951	1983	2011	
April	8.34	6.22	5.94	5.86	4.67	4.66	4.44	4.42	4.00e	3.94	
	1983	2011	2007	1996	1984	2005	1976	1973	1952	1958	
May	9.72	9.05	8.08	6.80	6.32	5.69	5.63	5.42	5.22	5.00e	
	1984	1983	2004	1989	1973	2009	2006	1972	2011	1945,52	
June	8.04	7.98	7.91	7.32	6.69	6.52	6.01	5.78	5.69	5.59	5.23
	2011	1972	1998	1987	2005	1989	1982	2000	2009	1986	2006
July	10.49	10.00	9.00e	6.83	6.63	5.94	5.59	5.58	5.57	4.69	4.64
	2008	1996	1945	2004	1976	1988	1960	2009	1984	2005	1989
August	9.45	7.84	7.16	7.05	6.75	6.64	6.10	5.96	5.93	5.49	5.31
	2011	1990	1975	1983	1988	1955	1971	2003	1997	2004	2009
September	8.94	8.15	7.96	7.50	7.31	6.03	5.93	5.85	5.58	5.12	
	2011	1987	1999	2011	2003	1981	1977	1960	1975	1996	
October	8.61	8.53	8.23	6.88	6.83	6.29	6.18	6.10	5.71	5.53	5.00
	2005	2010	1995	1955	1976	1959	1987,90	2009	1977	1975	2007
November	8.02	6.44	6.40	5.69	5.00	4.99	4.76	4.66	4.50e	4.36	
	1972	1969	1983	1997	1959	1988	1963	1974	1954	1977	
December	8.00	7.37	6.97	6.68	5.89	5.85	5.37	4.99	4.85	4.73	4.24
	1948	1973	2008	1983	1996	1969	2003	2004	1990	1957	1981
Annual	63.87	57.37	55.53	53.23	52.77	52.69	52.42	51.93	50.14	50.00e	
	2011	2008	1983	1972	2006	2005	1990	1996	2003	1945	

TABLE 67
INDIAN LAKE 1899 – 2012
Wettest Months

	1	2	3	4	5	6	7	8	9	10
January	6.43 1979	6.16 1999	5.81 1998	5.59 1924	5.56 1947	5.28 1978	5.06 1932	4.79 1913	4.73 1930	3.56 1917
February	6.62 2008	6.15 1909	6.12 1981	5.65 1908	5.37 1910	5.11 1954	4.98 1918	4.95 1972	4.51 1951	3.76 1938
March	8.32 1913	8.12 2011	6.22 1955	6.08 1925	6.02 1900	6.0e 1899	5.88 2008	5.65 1980	5.30 1951	4.83 1903
April	8.44 2011	6.62 1945	6.12 2000	6.08 1929	5.79 1954	5.68 1902	5.51 1914	5.35 1983	5.13 1993	4.80 1951
May	8.68 2011	7.26 1947	7.18 1945	6.74 1990	6.50 2009	6.10 1908	5.73 2000	5.54 1973	5.52 1984	4.96 1953
June	10.31 1922	9.00 1905	8.96 1998	7.71 1944	7.62 1917	7.10 1947	6.75 1972	6.50 2009	5.49 1968	5.13 2011
July	8.31 1902	8.29 1947	7.78 1935	7.06 1931	7.00 1951	6.86 1932	6.42 2008	6.41 1986	6.27 1941	6.22 1995
August	6.82 1979	6.61 1964	6.46 1949	6.45 2010	6.33 1932	6.00e 2011	5.89 1977	5.68 1998	5.52 2004	5.40 1955
September	10.05 1942	8.82 1945	8.39 1938	7.59 1956	7.42 1985	6.87 1999	6.51 1989	6.11 1981	5.55 2011	5.10 1904
October	9.48 1995	8.34 2005	8.33 1917	8.12 1959	7.64 1945	7.60 1932	7.48 1955	6.91 2010	6.45 1990	6.32 2003
November	8.57 1927	8.02 1959	7.45 1950	7.16 1900	6.65 1920	5.0e 1989	4.66 1974	4.50 1963	4.29 1996	3.52 1987
December	7.0e 1983	6.50 1920	6.31 1952	6.06 1957	5.79 2008	5.77 1973	5.75 1907	4.32 1984	4.10 1942	3.81 2009
Annual	58.59 2011	55.30 1945	53.45 1976,08	52.15 1951	51.15 1972	51.03 1954	48.91 1947	45.96 2005	45.79 1990	48.78 1929

TABLE 68
LITTLE FALLS CITY RESERVOIR 1897 -99   1901-2011
Wettest Months / Years

	1	2	3	4	5	6	7	8	9	10	
January	5.18	4.94	4.90e	4.80	4.66	4.15	4.00e	3.95	3.68	3.52	
	1950	1978	1999	1932	1979	1910	2005	1987	1996	1958	
February	5.49	5.48	4.41	3.85	3.82	3.56	3.50	3.40	3.30	3.08	
	1909	1971	1950	1910	1915	1990	1926	1981	2008	1924	
March	6.09	5.21	5.15	5.14	5.00e	4.61	4.43	3.68	3.60	2.93	
	1913	1903	1994	1921	2001	1936	1977	1953	1993	1980	
April	9.22	7.50e	7.20	5.17	5.10e	5.00e	4.79	4.75	4.41	4.30	
	1983	2011	1929	1972	2011	2005	1945	1948	1990	1996	
May	8.02	7.49	7.21	6.32	6.30	6.12	5.91	5.66	5.59	5.15	
	1990	1973	1953	1989	1984	2004	1947	1901	1945	1929	
June	8.58	7.34	6.99	6.98	6.82	6.73	5.78	5.67	5.36	5.03	
	1972	1989	1928	1903	2010	2006	1947	1944	1981	1945	
July	9.07	7.94	7.23	6.73	6.30	5.96	5.48	5.44	5.03	4.98	4.91
	1902	1971	1915	1987	2004	1945	2010	1927	1988	1948	2003
August	7.63	7.25	6.87	6.76	6.55	6.44	6.36	6.00e	5.84	5.61	
	1936	1903	1950	2009	2010	2004	1964	2011	1983	1990	
September	9.66	8.86	8.20	7.16	6.80e	6.00e	6.29	5.96	5.21	5.13	
	1977	1938	1945	2010	1999	2011	1950	1981	1960	1987	
October	8.49	7.58	7.26	7.24	6.76	6.52	6.04	5.87	5.80	5.39	
	2005	1903	1976	1995	1932	1955	1936	1945	2006	2010	
November	7.99	7.48	6.16	6.02	5.94	5.38	5.21	5.06	4.85	4.56	
	1927	1972	1921	1977	1954	1932	1945	1988	1995	2007	
December	7.00e	6.80	6.07	5.60	5.35	5.29	5.23	4.83	4.64	4.61	
	2010	1973	1972	1969	1983	1996	1901	1948	1927	1990	
Annual	55.08	53.37	51.69	50.52	48.43	48.05e	46.83	47.41	45.87	45.00	
	1950	1990	1945	1972	1977	2003	1975	1996	2006	2010	

Missing 1999 & 2000

TABLE 69
WHITE PLAINS 1864 - 1892
WESTCHESTER AIRPORT 1946 - 2011
Wettest Days

	1	2	3	4	5	6	7	8	9	10	
January	5.50	4.70	3.29	3.21	3.17	2.97	2.90	2.88	2.80	2.64	
	1874	1882	1979	1881	1999	1982	1878	1888	1883	1997	
February	4.70	3.90	3.85	3.25	2.77	2.75	2.70	2.66	2.50	2.31	
	1882	1888	1880	1886	1973	2.75	1969	1970	1883	1981	
March	6.10	3.79	3.52	3.23	3.15	3.12	3.10	3.00	2.83	2.81	
	1876	1951	1953	2007	1980	1888	1977	1882	1979	1962	
April	6.82	4.46	3.80	3.13	3.10	2.99	2.92	2.85	2.80	2.77	
	2007	1984	1952	2011	1876	1987	2006	1885	1889	1983	
May	4.80	4.20	3.34	2.88	2.81	2.62	2.58	2.56	2.33	2.32	
	1968	1989	2006	1992	2002	1984	1976	1952	1886	1946	
June	6.40	5.40	3.30	2.93	2.92	2.84	2.69	2.62	2.56	2.49	
	1884	1972	1892	1997	2000	2005	2001	1890	1984	1969	
July	5.90	3.50	3.20	3.19	3.04	3.02	3.01	3.00	2.99	2.74	
	1875	1884	2000	1960	1980	1884	2000	1880	1984	1965	
August	5.18	3.90	3.57	3.53	3.50	3.20	3.04	3.02	2.96	2.94	2.91
	2011	1879	1954	1971	1955	1874	2010	1884	1962	1985	1991
September	5.93	5.90	5.06	5.01	4.93	4.90	4.70	4.42	4.18	4.09	
	2004	1874	1975	1974	1954	1999	1878	2008	1966	1971	
October	9.70	5.30	5.09	4.77	4.34	4.12	4.05	3.97	3.84	3.78	
	1877	1877	1983	1955	2005	1989	1890	1996	2005	1995	
November	5.04	4.95	3.90	3.51	3.46	3.29	3.09	2.94	2.68	2.44	
	1889	1977	1878	1972	2006	1950	1955	1990	1954	1962	
December	4.45	4.10	3.35	2.83	2.80	2.60	2.46	2.42	2.36	2.30	
	1879	1878	1888	1968	1887	1882	1891	1993	1992	1952	

TABLE 70
CARMEL / YORKTOWN HEIGHTS 1888 – 2012
Wettest Days
Inches

	1	2	3	4	5	6	7	8	9	10
January	3.33	2.63	2.57	2.55	2.52	2.45	2.35	2.20	2.12	2.08
	1979	1910	1940	1905	1996	1888	1909	1979	1986	1936
February	3.30	2.82	2.70	2.56	2.44	2.42	2.24	2.19	2.14	2.12
	1973	1941	1915	1896	1898	1958	1900	1977	1893	2003
March	4.60	3.44	3.25	3.20	3.06	2.97	2.95	2.86	2.85	2.65
	1980	1919	2011	2011	1977	1903	1987	2010	1913	2007
April	5.15	3.60	3.52	3.40	3.00	2.98	2.97	2.89	2.82	2.63
	2007	1987	2005	1901	2005	1980	2011	1984	1968	1924
May	5.57	3.75	3.17	2.90	2.47	2.44	2.27	2.26	2.24	2.20
	1900	1989	1984	1979	1927	1922	1937	1978	1947	1999
June	4.95	4.30	3.40	3.06	3.05	3.04	2.82	2.55	2.50	2.48
	2001	1969	1972	1994	1973	1969	2000	2011	1903	2005
July	5.03	4.24	4.22	4.02	3.75	3.50	3.44	3.35	3.34	3.31
	1897	1946	2012	1928	1984	1941	1938	1973	1996	1947
August	6.86	6.35	6.00	5.20	4.93	4.90	4.33	4.32	4.04	3.56
	1971	2011	1955	1965	1950	1952	1990	1961	2005	1926
September	11.00	5.10	4.18	4.11	3.84	3.72	3.71	3.50	3.31	3.25
	1999	2004	1937	1954	2008	1975	1920	1907	1899	1977
October	6.77	6.53	4.98	4.64	4.37	3.45	3.25	3.24	3.20	3.08
	1955	1903	1996	2005	1923	1995	2010	1987	1907	2006
November	3.47	3.07	3.04	3.02	3.01	2.99	2.97	2.93	2.89	2.86
	1947	1903	1947	1942	1943	1924	1994	1988	1977	1956
December	3.40	3.29	3.19	3.00	2.90	2.87	2.83	2.80	2.22	2.10
	1901	1941	1948	2008	1986	1895	1957	1983	1902	1903,90

TABLE 71
PEEKSKILL
1890, 91 1946 – 2012
Wettest Days of Record
Inches

	1	2	3	4	5	6	7	8	9	10
January	3.70	2.74	2.57	2.25	2.20	2.18	2.10	2.05	2.00	1.92
	1980	1979	1979	1984	2003	1999	1986	1891	1994	2010
February	3.40	2.55	2.50	2.38	2.10	2.01	1.95	1.89	1.75	1.73
	1973	1890	1981	2003	1983	1966	1969	1977	1988	1958
March	3.25	3.22	3.00e	2.95	2.58	2.44	2.24	2.22	2.08	2.06
	1951	1980	2011	1987	1979	2010	1997	2011	1953	1953
April	4.02	3.22	3.13	2.40	2.34	2.28	2.21	2.09	2.06	2.05
	1984	1980	1987	1968	1970	1990	1983	1976	1952	1980,83
May	3.40	3.29	3.15	2.45	2.20	2.20	2.08	2.06	2.04	2.00
	1968	1984	1984	1989	1992	1981	1976	1952	1978	1979
June	3.34	3.02	2.95	2.52	2.51	2.37	2.33	2.30	2.20	2.12
	2001	2001	1969	1960	2000	1952	1973	1973	1996	2001
July	3.42	3.10	3.04	3.00	2.81	2.55	2.44	2.43	2.18	2.17
	1946	2004	1952	1996	1960	1953	1960	1984	1986	1948
August	6.16	5.61	5.33	5.19	4.91	4.90	4.05	3.75	3.44	3.21
	1926	2011	1961	1952	1955	1952	1990	1971	1965	1971
September	9.00	4.30	4.27	4.17	3.90	3.60	3.47	3.33	3.18	3.12
	1999	1991	2004	1960	2004	1966	1977	2008	1954	1985
October	5.56	4.40	3.70	3.55	3.30	3.13	3.10	2.97	2.90	2.88
	2005	1955	1995	2005	1995	1989	1995	1966	1996	2010
November	4.08	3.42	3.02	3.00	2.82	2.79	2.71	2.65	2.64	2.62
	1977	1990	1993	1945	1988	1946	1950	1954	1947	1999
December	2.98	2.88	2.58	2.36	2.07	2.01	2.00	1.88	1.85	1.83
	1948	1957	1983	2003	2011	1968	1947	1953	1957	2000

TABLE 72
WEST POINT 1890 –2012
Wettest Days
Inches

	1	2	3	4	5	6	7	8	9	10
January	3.78	3.40	2.82	2.70	2.50	2.27	2.26	1.99	1.93	1.89
	1996	1978	1996	1979	1921	1986	1979	1953	1940	1982
February	3.18	2.60	2.40	2.35	2.04	2.00	1.97	1.82	1.73	1.56
	1977	1898	1973	1941	1961	1925	1983	1983	1970	1936
March	4.40	4.09	3.90	3.80	3.70	3.39	3.29	2.32	2.18	2.17
	1951	2011	1987	1896	1896	1952	1999	2010	1958	1969
April	4.20	3.85	3.55	3.20	3.10	3.00	2.47	2.44	2.30	2.20
	1910	1952	1895	1910	1987	2005,11	1968	1970	1924	2002
May	3.98	3.38	3.37	2.89	2.60	2.58	2.35	2.22	2.13	2.10
	1984	1946	1981	1908	1908	1910	1989	1937	1984	1985,92
June	3.63	3.08	2.76	2.74	2.45	2.27	2.18	2.15	2.12	2.10
	1952	2000	1938	1973	1973	1959	1998	1986	1994	1994
July	3.75	3.70	3.40	3.29	3.21	3.16	3.00	2.86	2.85	2.38
	1996	1897	1922	1960	1942	1984	1994	1938	1928	1945
August	4.39	4.00	3.37	3.22	2.92	2.68	2.60	2.50	2.46	2.45
	2011	1971	2011	1955	1976	1971	1954	1895	1960	1955
September	8.00	6.11	5.50	4.76	4.56	4.00	3.95	3.74	3.60	2.99
	1999	2003	1924	1960	1954	1907	1938	1937	2004	1956
October	9.15	5.50	4.80	4.70	4.50	4.20	3.38	2.80	2.78	2.77
	2005	2010	1907	2005	1995	1955	1895	1987	1929	1983
November	4.41	4.00	3.80	3.78	3.30	3.29	2.96	2.90	2.84	2.80
	1927	1977	1896	1993	1990	1932	1979	2006	1969	2002
December	3.89	3.80	3.79	3.60	3.30	3.10	2.92	2.85	2.50	2.48
	1957	2008	1983	1983	1948	1909	1990	1952	1996	2003

TABLE 73
MIDDLETOWN 1893 – 1908 1951 – 2010
Wettest Days
Inches

	1	2	3	4	5	6	7	8	9	10	
January	2.15	2.10	2.00	1.93	1.90	1.88	1.72	1.54	1.49	1.48	
	1978	1979	1996	1905	1986	1953	1979	1898	1982	1953	
February	3.60	2.60	2.10	1.94	1.90	1.85	1.47	1.43	1.39	1.32	
	1896	2008	1898	1965	1986	1958	1978	1900	1975	1962	
March	2.75	2.60	2.45	2.36	2.34	2.20	2.04	1.90	1.86	1.68	1.66
	2011	1896	1999	1980	1905	2008	1901	2005	1978	1993	1983,91
April	4.00	3.60	2.82	2.68	2.40	2.35	2.30	1.90	1.88	1.80	1.67
	1895	2007	1997	1952	2005	1968	1983	1975	1984	1901	2011
May	3.42	2.51	2.50	2.47	2.40	2.30	2.25	2.23	2.10	2.05	
	1968	1981	1972	1894	1952	1904	1989	1984	1908	1986	
June	4.25	2.95	2.85	2.81	2.71	2.30	2.21	2.18	2.15	2.05	
	1952	1992	2009	1973	1967	1980	1998	1895	2009	1972	
July	5.51	4.06	3.50	3.02	3.00	2.70	2.50	2.48	2.47	2.25	
	1959	1963	1897	1984	1975	1906	1996	1897	1992	1996	
August	5.70	4.90	4.34	4.28	3.34	3.08	3.04	3.02	2.98	2.79	
	1955	1991	1955	1971	1960	1898	1893	1902	1901	1990	
September	6.00	4.87	3.64	3.10	3.04	3.00	2.98	2.90	2.78	2.67	
	1894	1952	1904	1999	1979	1989	2004	2003	1997	1977	
October	8.00	4.56	4.12	4.02	2.97	2.91	2.70	2.69	2.50	2.35	
	1903	1959	1995	2010	1904	1895	1895	1955	2002	1955	
November	3.64	2.93	2.80	2.72	2.50	1.94	1.84	1.82	1.78	2.03	
	1895	1977	1897	1907	1954	1985	2003	1987	2006	2003	
December	2.87	2.75	2.70	2.20	2.15	2.10	1.94	1.93	1.70	1.63	
	1990	1953	2008	1897	2000	1957	1993	1983	1900	1975	

TABLE 74
MOHONK LAKE 1891-2012
Wettest Days
Inches

	1	2	3	4	5	6	7	8	9	10
January	2.91	2.25	2.22	2.12	2.07	2.02	2.00	1.90	1.87	1.85
	1979	1986	1978	1982	1979	1902	1924	2006	1979	1915
February	4.30	3.90	3.40	2.63	2.91	2.40	2.14	2.05	2.00	1.80
	1898	1896	1920	1900	1926	1914	1958	1953	1916,21	1912
March	6.50	4.70	3.00	2.90	2.82	2.77	2.67	2.62	2.39	2.35
	1914	1896	2008	1899	1951	2008	1995	1999	2010	1913
April	4.13	3.61	3.44	3.11	3.07	3.00	2.85	2.50	2.48	2.30
	1987	2007	1910	1952	1970	1901	1910	1902	1924	1928
May	4.16	4.10	3.27	3.20	3.19	2.90	2.73	2.40	2.26	2.25
	1968	1937	1992	1989	1981	1999	1908	1998	1984	1947
June	3.56	3.44	3.35	3.17	3.13	2.78	2.51	2.46	2.40	2.18
	1952	1934	1938	1973	1903	1932	2009	1961	2006	1982
July	5.06	4.05	3.92	3.85	3.24	3.23	2.84	2.82	2.80	2.78
	1996	1902	1960	1923	1934	1897	1958	1952	2000	1897
August	8.21	6.35	5.04	3.54	3.50	3.49	3.09	3.06	3.03	3.00
	2011	1955	1971	1910	1901	1955	1954	1933	1994	1900
September	4.83	4.40	4.00	3.83	3.80	3.79	3.75	3.74	3.70	3.10
	1960	1904	1999	1985	1938	1979	1938	1987	2003	2006
October	6.16	5.33	5.07	4.50	3.68	3.61	3.40	3.30	3.16	3.14
	2005	1955	1903	1995	2006	2010	1911	2002	1955	1996
November	5.00	4.55	3.28	3.15	2.98	2.80	2.40	2.30	2.27	2.25
	1927	1977	1896	1943	1932	1897	1932	2005	1942	1919
December	4.03	3.80	3.70	3.30	2.60	2.57	2.40	2.38	2.37	2.33
	1948	2000	1915	1990	1953	1983	1973	1915	1945	1915, 57

TABLE 75
WETTEST DAYS OF RECORD
Albany City & Airport
1895 -2012
Inches

	1	2	3	4	5	6	7	8	9	10	
January	1.78	1.72	1.68	1.65	1.60	1.54	1.37	1.33	1.27	1.25	
	1986	2006	1990	1996	1978	1936	1969	1979	1983	1976	
February	1.89	1.84	1.66	1.60	1.51	1.50	1.48	1.47	1.46	1.39	
	1896	1941	1920	1990	1926	1914	1895	1977	1975	1993	
March	2.14	2.02	2.00	1.98	1.74	1.73	1.71	1.70	1.66	1.64	
	1900	1997	1999	1977	2008	1932	1986	1990	1901	2005	
April	2.60	2.26	1.91	1.90	1.80	1.76	1.75	1.68	1.65	1.59	
	1968	2007	1970	1975	1996	1983	1945	1993	1983	2006	
May	2.25	2.16	2.03	2.02	2.01	1.90	1.80	1.71	1.67	1.62	
	1984	2006	1906	1938	1931	1984	1897	1954	1998	1979	
June	3.30	3.26	2.64	2.51	2.45	2.34	2.30	2.28	2.22	2.13	
	2000	1952	1934	1922	1976	1973	1904	1944	2006	1903	
July	3.49	3.23	3.11	2.95	2.90	2.82	2.76	2.52	2.50	2.34	
	1996	2000	1897	1915	1960	1980	2009	1958	1969	1945	
August	4.69	4.23	3.50	3.24	2.90	2.82	2.78	2.68	2.27	2.01	1.95
	2011	1927	1971	1960	1954	1901	1980	1950	1936	1949	2004
September	5.70	3.58	3.25	3.00	2.76	2.68	2.62	2.51	2.35	2.33	
	1999	1960	1938	1966	1916	2010	1934	1920	1924	1950	
October	3.09	3.04	2.87	2.82	2.65	2.63	2.62	2.31	2.30	2.22	
	1903	2010	1932	1987	1995	2005	1972	1911	1998	1904	
November	3.28	2.32	2.23	2.17	1.85	1.79	1.76	1.73	1.68	1.67	
	1927	1959	1924	1991	1920	1972	1895	1991	1993	1918	
December	3.19	2.91	2.90	2.67	1.98	1.86	1.82	1.79	1.67	1.61	
	1952	1948	2000	1948	1915	1983	1950	2002	1973	1920	

TABLE 76
TROY 1913-35 1956-2012
Wettest Days
Inches

	1	2	3	4	5	6	7	8	9	10	
January	1.80	1.65	1.60	1.50	1.48	1.40	1.35	1.33	1.30	1.22	
	2001	1990	1961	2008	2006	1995	1983	1964	1976	1925,78	
February	1.58	1.40	1.35	1.27	1.25	1.20	1.19	1.16	1.11	1.10	
	1934	1977	1981	1929	1914	2008	1929	1990	1920	2008	
March	3.10	2.81	2.10	2.00	1.90	1.85	1.82	1.80	1.78	1.75	
	2001	1986	2008	1977	2005	1984	2007	2001	1919	2011	
April	2.67	2.30	2.15	1.95	1.76	1.73	1.70	1.61	1.60	1.48	
	1968	2007	1990	1983	1991	1993	1998	1924	1977	1973	
May	2.40	2.20	2.03	1.91	1.90	1.85	1.70	1.65	1.61	1.60	
	1984	2004	1984	1983	1996	1931	2002	1979	2007	1922	
June	3.80	2.84	2.68	2.30	2.25	2.23	2.10	2.02	1.92	1.88	
	1973	1934	2007	2006	2000	1922	2002	1982	1996	1998	
July	2.88	2.70	2.65	2.60	2.52	2.50	2.34	2.32	2.30	2.28	
	1960	2009	1976	2008	2005	2004	2006	1966	1996	1968	
August	4.00e	3.35	2.82	2.71	2.56	2.50	2.30	2.27	2.15	2.12	2.06
	2011	1971	1975	1986	2001	2000	1961	1933	1927	1926	1982
September	3.00	2.75	2.60	2.59	2.40	2.25	2.23	2.14	2.11	2.10	
	1987	1960	2003	1920	1960	1983	2004	1933	1933	1999	
October	4.40	2.53	2.30	2.20	2.15	2.06	2.02	2.00	1.98	1.96	
	2010	1932	1995	2005	1995	1927	1972	1998	1912	1987	
November	2.62	2.40	2.15	2.00	1.99	1.88	1.82	1.80	1.56	1.55	
	1990	1927	1969	1995	1959	1924	1932	2005	1927	2003	
December	2.50	1.90	1.79	1.70	1.68	1.53	1.50	1.30	1.28	1.25	
	2000	1957	1983	1914	2002	2003	2008	2003	2005	1973	

TABLE 77
SARATOGA SPRINGS 1895 - 1906 1941 - 2012
Wettest Days
Inches

	1	2	3	4	5	6	7	8	9	10	
January	2.50	1.92	1.91	1.80	1.46	1.40	1.38	1.36	1.33	1.31	
	1998	1986	1979	2000	1990	1999	1991	1898	1976	2006	
February	2.10	2.09	1.90	1.58	1.51	1.47	1.45	1.43	1.40	1.35	
	1898	1945	1958	1971	1982	1965	1977	1900	1972	1946	
March	3.10	3.08	2.21	2.10	2.08	2.00	1.61	1.58	1.56	1.55	
	2005	1977	1993	2008	2005	2002	1997	1986	1896	1990	
April	3.10	2.20	2.10	2.05	1.95	1.65	1.60	1.50	1.48	1.36	
	2006	1970	2007	1975	1968	1977	2006	2000	1958	1976	
May	2.70	2.53	2.21	1.80	1.74	1.70	1.66	1.60	1.57	1.50	
	2004	1984	1992	2000	1979	1983	1989	1972	1947	1998, 02	
June	3.39	3.30	2.60	2.23	2.17	2.16	1.95	1.84	1.81	1.74	
	1973	2006	2002	1972	2005	1982	1949	1987	1975	1987	
July	3.05	2.51	2.50	2.48	2.47	2.30	2.21	2.20	2.02	2.01	
	1994	1994	2008	2000	1974	2001	1996	1984	1944	1962	
August	4.45	4.23	3.74	3.28	3.27	2.57	2.49	2.45	2.43	2.16	2.14
	1941	2011	1971	1988	1986	1975	1950	1982	1990	1955	1995
September	5.49	4.00	2.80	2.48	2.33	2.23	2.03	2.02	2.00	1.98	
	1899	1999	1960	1987	1966	1950	1987	1979	2004	2003	
October	4.74	2.80	2.30	2.29	2.28	2.21	2.20	2.08	2.04	2.02	
	2010	1972	1976	1987	1943	1898	1977	1990	2005	1975	
November	2.65	2.05	2.00	1.99	1.98	1.92	1.88	1.79	1.65	1.63	
	1972	1990	2003	1959	1947	1996	1943	1898	1948	2007	
December	3.52	2.94	2.20	2.00	1.96	1.80	1.75	1.74	1.73	1.70	
	1948	1948	2000	1969	1983	2002	2004	1994	1967	2008	

TABLE 78

GLENS FALLS City & Airport 1893 - 2012
Wettest Days of Record
Inches

	1	2	3	4	5	6	7	8	9	10
January	2.84	2.70	2.40	2.01	1.84	1.80	1.70	1.66	1.63	1.58
	2006	1999	1998	1986	1978	2000	1925	1983	1979	1994
February	2.70	2.61	2.50	2.38	2.26	2.24	2.03	2.00	1.99	1.89
	1899	1909	2006	1981	1915	1914	1900	1954	1951	1974
March	3.36	2.50	3.00e	2.20	2.18	2.03	2.02	2.00	1.92	1.85
	1994	1977	2011	2008	1896	1900	1964	2010	1942	1903
April	2.80	2.66	2.54	2.40	2.24	2.15	2.12	2.10	2.04	2.00
	2007	1968	1895	1970	1975	1970	1962	1952	1976	1996
May	2.52	2.34	2.30	2.27	2.11	2.07	2.06	1.90	1.73	1.72
	1952	1989	1992	1984	1966	1977	1906	1981	1989	1983
June	4.21	3.48	3.44	3.40	3.36	2.74	2.60	2.58	2.48	2.10
	1944	1973	1987	2006	1952	1976	1904	1903	1958	1944,52
July	4.59	3.20	3.08	3.01	2.84	2.80	2.70	2.67	2.60	2.53
	1919	2006	1957	1944	1988	1997	2000	1978	1915	1908
August	4.52	3.67	3.65	3.62	3.10	3.08	2.92	2.89	2.83	2.78
	1949	2011	1971	1988	1990	1995	2011	1984	1924	1954
September	4.60	4.48	3.75	3.70	3.33	3.22	3.02	2.76	2.68	2.65
	1999	2010	1979	1999	1899	1966	1960	1987	1981	2003
October	3.57	2.80	2.78	2.70	2.63	2.60	2.58	2.57	2.48	2.41
	1972	2005	1955	1976	1902	1990	1927	2003	1911	2010
November	3.12	3.10	3.05	2.98	2.76	2.38	2.08	2.00	1.98	1.96
	1943	1927	1896	1897	2003	1959	1990	1947	1932	1912
December	3.10	2.91	2.90	2.61	2.50	2.33	2.20	2.18	2.10	2.00
	1948	1948	1996	1952	2000	1920	2006	1973	1953	1894

TABLE 79
INDIAN LAKE 1899 – 2012
Wettest Days
Inches

	1	2	3	4	5	6	7	8	9	10	
January	2.35	2.20	2.05	1.90	1.85	1.71	1.70	1.64	1.62	1.60	
	1924	1976	1959	1925	1923	1957	2006	1998	1980	1978	
February	2.78	2.24	2.10	2.05	2.00	1.72	1.63	1.60	1.58	1.56	
	1954	2000	1981	1908	2007	1951	1982	1951	1941	1998	
March	2.65	2.40	2.04e	2.00	1.96	1.95	1.93	1.84	1.80	1.78	
.	1900	1994	2011	1913	1977	1925	1993	1985	1947	1903	
April	2.20	2.15	2.01	1.72	1.63	1.60	1.55	1.54	1.53	1.50	1.47
	1913	1902	1968	2011	1976	1904	2006	1920	2002	1908	1992
May	2.32	2.23	2.06	2.04	2.03	2.00	1.83	1.80	1.70	1.60	
	1945	1969	1916	1906	1991	2009	1937	2002	1916	2000	
June	3.25	3.20	3.14	3.05	3.00	2.80	2.14	2.13	2.10	2.08	
	1947	1944	1942	1940	1987	2006	1973	1944	1919	1922	
July	2.42	2.39	2.32	2.30	2.25	2.10	2.08	1.90	1.85	1.80	
	2000	1937	1947	1900	1958	1990	1902	2006	1996	1926, 86	
August	3.74	3.05	3.03	2.61	2.60	2.50	2.48	2.42	2.40	2.15	2.10
	2011	1964	1949	1966	1998	2005	2003	1971	1933	1904	1910,15
September	3.72	3.50	3.30	3.25	2.60	2.57	2.56	2.50	2.40	2.37	
	1956	1999	1975	1924	1941	1945	1945	1942	1938	1985	
October	4.40	3.34	3.30	2.87	2.75	2.60	2.30	2.28	2.27	2.20	
	1932	2010	1988	1904	1945	2005	1981	1955	1903	1995,05	
November	3.24	2.90	2.78	2.75	2.31	2.14	2.01	1.86	1.85	1.80	
	1959	1996	1950	1920	1967	1948	1982	1948	1899	1995	
December	2.76	2.72	2.60	2.05	2.02	1.96	1.70	1.58	1.55	1.50	
	1952	1948	1900	1927	1920	1957	1990	1952	1986	1907,29	

TABLE 80
EXTREME DAILY PRECIPITATION (Inches)
WITH MONTH OF OCCURENCE

HUDSON VALLEY DIVISION

	Jan	Feb	Mar	Apr	May	Jun	Jul	Aug	Sep	Oct	Nov	Dec
Albany City	1.78	1.89	2.14	2.60	2.25	3.30	3.49	4.23	5.70	3.09	3.28	3.19
& AP 1895-	1986	1896	1900	1968	1984	2000	1996	1927	1999	1903	1927	1952
Glens Falls	2.70	1.89	3.20e	2.66	2.30	4.21	4.59	3.65	3.70	3.57	2.08	2.91
Airport	1999	1974	2011	1968	1992	1944	1919	1971	1999	1972	1990	1948
Glens Falls	2.84	2.61	3.90	2.80	2.52	3.48	3.84	4.52	4.50	2.80	3.10	3.70
Farm	2006	1909	2011	2007	1952	1973	1919	1949	1999	2005	1927	1948
Middletown	2.15	3.70	2.75	3.60	3.42	4.25	5.51	5.70	6.00	4.56	3.64	2.87
	1978	1896	2011	2007	1968	1952	1959	1955	1894	1959	1895	1990
Mohonk Lake	2.91	4.30	4.70	4.13	4.16	3.56	5.06	8.21	4.83	6.20	4.55	4.03
	1979	1898	1896	1987	1968	1952	1996	2011	1960	2005	1977	1948
Poughkeepsie	2.02	2.17	2.22	4.20	2.80	2.84	4.72	5.50	6.05	5.70	3.36	3.22
City &AP	1979	1977	1977	2007	1968	1973	1975	1955	1950	2005	1932	1948
Saratoga	2.50	2.09	3.10	3.20	2.70	3.39	3.00	4.45	5.49	2.80	2.65	3.52
Springs	1998	1945	1977	2006	2004	1973	1994	1941	1899	1932	1972	1948
Troy	1.80	1.58	3.10	2.67	2.40	3.80	2.88	3.35	3.00	2.53	2.62	2.50
	2001	1934	2001	1968	1984	1973	1960	1971	1987	1932	1990	2000
West Point	3.78	3.18	4.40	4.10	3.98	3.63	3.75	4.00	8.00	9.15	4.15	3.89
	1996	1977	1951	1910	1984	1952	1996	1971	1999	2005	1927	1957
Yorktown	3.33	3.37	4.60	5.15	5.57	4.95	5.03	6.86	11.00	6.77	3.47	3.29
Heights	1979	1900	1980	2007	1900	2001	1897	1971	1999	1955	1947	1941
Slide Mountain	5.34	3.50	3.30	6.64	4.26	4.92	6.62	6.00	6.00	8.24	7.78	4.80
	1976	1981	1999	1987	1981	1972	1952	1955	1999	1955	1950	1996

MOHAWK VALLEY DIVISION

	Jan	Feb	Mar	Apr	May	Jun	Jul	Aug	Sep	Oct	Nov	Dec
Gloversville	2.00	2.42	2.80	3.30	2.31	3.10	2.72	3.50	3.02	3.89	2.58	2.79
	2001	1990	1977	2006	1952	2006	1932	2000	1999	1932	1935	1948
Little Falls	2.00	1.70	2.00	2.05	2.16	3.22	3.39	3.09	2.52	2.98	3.50	1.85
City Reservoir	1925	1998	1977	1983	1918	2008	2006	1949	1905	1932	1996	1942
Utica	2.20	2.38	2.31	2.28	2.75	3.00	6.06	3.05	4.14	3.06	2.40	2.16
	1925	1971	1994	1977	2000	2006	1965	1975	1985	1955	1996	1973

NORTHERN PLATEAU DIVISION

	Jan	Feb	Mar	Apr	May	Jun	Jul	Aug	Sep	Oct	Nov	Dec
Indian Lake	2.34	2.78	2.40	2.20	2.32	3.25	2.42	3.05	3.72	4.40	3.24	2.76
	1924	1954	1994	1913	1945	1947	2000	1964	1956	1932	1959	1952
Newcomb	2.05	3.02	2.06	1.90	2.03	2.82	2.71	2.78	3.96	2.46	3.80	3.12
	1998	1954	1985	1970	2006	1987	2000	1989	1985	1992	1996	1948
North Creek	2.08	2.42	2.62	2.70	2.86	3.10	2.60	3.24	4.12	3.00	2.49	4.00
	1978	2007	1973	1969	2009	2006	1976	1971	1956	1995	1990	1948

An examination of the wettest days at seventeen stations shows that all but two months have had four or five days within the 1981 – 2010 period, The years 1948, 1955 and 1999 have had their heaviest amounts at the most stations. This heavy one day precipitation is another indication that along with the increasing warmth for this region at least has come an increase in heavy short term amounts.

STATION HISTORIES

ALCOVE DAM

Located in southern Albany County at an elevation of 601 feet above sea level, about 1000 feet east of the Alcove reservoir, daily temperatures have been recorded by employees of the Albany Water Department since 1948.

GLENS FALLS FARM / AIRPORT

Initial daily temperature readings were first taken by C.L. Williams in 1892 and continued to 1933 and then from 1942 to 1944. The thermometers were located one half mile north of the Post Office at the fire station in Warren County at an elevation of 330 feet. After a four year break temperature readings resumed in 1948 at the Glens Falls Water Department Farm 3.5 miles WNW of the Glens Falls Post Office at an elevation of 504 feet. Present day readings are taken by employees of the City Water Department.

Temperature readings at the airport located in Warren County at an elevation of 320 feet above sea level began in 1946. Airport employees monitor the station. The elevation differences between the two stations do result in daily and monthly temperature differences.

PEEKSKILL

One of the few reliable weather stations not affiliated with the National Weather Service Cooperative network, Peekskill with its unpublished data is maintained by employees of the Water Department who have taken daily temperature readings since 1946. Rainfall has been monitored since 1926. The station is located about one mile northeast of downtown Peekskill at an elevation of 370 feet above sea level. Its readings and averages are shown to be representative of the suburban region outside of Peekskill in Westchester County. While the instruments used are not the same as those of the Weather Service, this does not appear to have shown any climate effects different from other stations in the area.

CARMEL / YORKTOWN HEIGHTS

Following the termination of daily temperature readings in 1982, the monthly mean temperature record for Carmel dating back to 1888 was adjusted by a mean difference method to that of Yorktown Heights using data from the 18 year period of concurrent observations. This has resulted in the retention of a continuous record to the present, the longest in Westchester and Putnam. The station which is part of the U.S. Historical Climate network also monitors daily rain and snow.

The Carmel location, 1.2 miles southwest of the Carmel post office at an elevation of 490 feet above sea level was perhaps the best non urbanized site possible for any weather station in southeastern New York State. It was located on land near the West Branch reservoir of the

Croton watershed and was maintained by employees of the New York City Department of Water Supply. Daily fall low temperatures there were about 2 – 4 degrees higher than the local area and about that amount lower in the summer due to the influence of the nearby 1000 acre reservoir. Precipitation is still being monitored.

The Yorktown Heights component station of the combined temperature record has been in operation since June 1964 by Dr. Jerome S. Thaler. Its location cannot be considered typical of northern Westchester. Located on the northwest slope of French Hill in a one acre zoned residential division at an elevation of 670 feet, it is the highest of any past or present station in both counties. The high elevation has many daily minimum temperatures 2 – 5 degrees higher than the areas that are 100 – 200 feet lower in elevation, which includes most of the town of Yorktown. Generally the warmer minimums occur on many autumn, winter and spring nights after cold air masses have passed through and inversion conditions prevail. The daily maximum temperatures are 1 – 3 degrees cooler than the other parts of the town due to greater air movements at the higher elevation. These two temperature effects do not balance each other and the averages are cooler than most of the region as is shown by greater snow amounts and greater duration of snow on the ground. One can say that this station's temperature profile is representative of areas of Putnam and Westchester with elevations greater than 600 feet above sea level.

MOHONK LAKE

In January 1891 daily readings first began at Lake Minnewaska, elevation 1800 feet above sea level five miles southwest of Mohonk Lake. In 1896 the station was relocated to its present site in Ulster County at an elevation of 1245 feet near the mountain top lake and resort of the same name. Albert (to 1942) and then Daniel Smiley (1943 to 1988) were the cooperative observers for nearly 100 years. Since then Paul Huth and assistants have continued the uninterrupted family tradition of recording the three daily temperature readings and any precipitation. By virtue of its extraordinary unbroken record and only one site location since 1896, it is part of the U.S. Historical Climatology Network.

SARATOGA SPRINGS

Daily readings of temperature and precipitation began in 1895 and continued for eleven years. The observer and exact location for that period are unknown. Readings by members of the Fur Animal Experimental Station were not resumed until 1941 at 3 or 4 miles northwest of the Saratoga Springs Post Office at an elevation of 550 feet. A four year break in the record beginning in 1952 was followed in 1956 by Cooperative Observer Leroy Post who took daily readings 7 miles southeast of the post office at an elevation of 320 feet. In 1960 employees of the New York State Conservation Department took over daily readings at their location 4 miles southwest of the post office at an elevation of 310 feet. This site for the instruments has continued through 2008. At 1990 the daily readings were taken by employees of the New York State Department of Environmental Conservation. In view of the many site changes only data from the period after 1960 can be considered to be reliable.

TROY (LOCK & DAM)

While intermittent temperature readings were taken as early as 1826 through 1846, at the nearby Lansingburgh Academy for the N.Y. State Board of Regents, consistent

daily readings by employees of the Troy Times in Rensselaer county did not start till 1916 and continued to 1934 at an elevation of 35 feet. Daily readings including precipitation, but no snowfall were resumed in 1956 at the Lock and Dam near the Hudson River at an elevation of 24 feet. Employees of the U.S. Corps of Engineers take the daily readings at present.

POUGHKEEPSIE CITY / AIRPORT

The Poughkeepsie climate record in Dutchess county has had a long but fragmented history. From 1829 through 1849 daily readings were taken by school officials for the N.Y. State Board of Regents. From 1890 to 1900 Vassar College monitored temperature and precipitation.

Daily readings by employees of the Central Hudson Gas and Electric Corporation were resumed in 1928 and continued till 1970, at a site 1.5 miles south of the Post Office at an elevation of 100 feet above sea level. In 1971 employees of the city Department of Public Works (Water Purification) took readings to 1980 1 mile north of the Post Office at an elevation of 50 feet. Concurrent temperature readings with the airport through 1980 began in 1947. ASOS continuous temperature readings (unpublished after 1992) are taken there at an elevation of 154 feet above sea level. Needless to say there are climate differences at all these sites caused by the equipment and elevation changes. In 1994 another local station manned by cooperative observer Edward Summerfield began observations (elevation 170 feet) which have continued with breaks to the present.

UTICA

Sponsored and equipped by the N.Y State Board of Regents daily temperature readings began in 1826 and continued through 1853. From 1854 through 1857 readings were taken by the Utica Lunatic Asylum. From 1870 through 1886 the Smithsonian Institute sponsored daily observers. From 1887 through 1889 the Signal Corps of the U.S. Army assumed the responsibility. From 1889 to 1893, Thomas Birt recorded daily temperature and precipitation. Temperature readings resumed in 1927 by employees of the Utica Gas and Electric Corp. Between 1945 and 1951 there were 3 station relocations.   Concurrent published readings by employees of Utica Airport and City Southern Reservoir began in 1948. Station elevations have ranged from 410 feet to the present airport's 714 feet. Three cooperative observers took readings through 1996 when that station was terminated. Airport readings in Oneida County have continued with breaks to the present time.

LITTLE FALLS (CITY RESERVOIR)

Daily readings for over 100 years have been taken by the City Engineer and other employees at the city reservoir. Since its beginning in January 1897, the thermometer shelter has been sited at only one location 890 feet above sea level about 100 feet from the reservoir .3 miles southwest of the city post office in Herkimer County. This fact makes its long term record invaluable and the station is included in the U.S. Historical Climatology Network despite the missing years of 1999 and 2000

GLOVERSVILLE

Located in the Mohawk River (a tributary of the Hudson River) Valley watershed in Fulton County this station has been operated by at least eight documented volunteer cooperative observers since its inception in 1892. The longest period of observations by one individual was 20 years. There have been three site relocations of the thermometer shelter ranging from 770 to 850 feet above sea level and within one mile of the city post office. A two year break in observations occurred in 1991 and 1992 and in 2006, with no readings since 2008

MIDDLETOWN

From 1893 to 1908 daily readings were taken by two volunteer observers. A 54 year break followed and readings were restarted in 1951 by the Community Broadcasting Corporation for 3 years at an elevation of 670 feet. The station was then moved .7 miles northwest to its present location. Employees of the city water department now monitor the station at their filter plant located 2 miles northwest of the city post office in Orange County at an elevation of 700 feet above sea level.

INDIAN LAKE

The oldest currently operating U.S. Weather Service Cooperative station in the Northern Plateau climate division of the watershed in Hamilton County, this station also has the distinction of having an unbroken record since its inception in 1899 near the Indian Lake Dam. Part of the U.S. Historical Climatology Network it has been serviced by a succession of ten volunteer cooperative observers monitoring temperature and precipitation. It is located 2 miles southwest of the post office at an elevation of 1660 feet above sea level. Daily readings have been taken by Darrin Harr since 2002.

NEWCOMB

One of only two Adirondack weather stations located within the Hudson Valley watershed in Essex County, daily temperature monitoring began in 1940. Since then there have been four site changes all within four miles of Newcomb at elevations between 1600 and 1700 feet above sea level. From 1978 through 1984 only precipitation was measured. Since its inception to 2004 there have been six volunteer cooperative observers. Since 2005 employees of the Newcomb Visitor Interpretation Center have continued daily readings.

WEST POINT

This long-term weather station, part of the U.S. Historical Climatology Network has the distinction of being one of the oldest active in the United States. The earliest temperature readings began in 1819 at the Post Hospital of the fledgling military academy in Orange County. Readings then were taken each day from one thermometer at 7 AM, 2 PM and 9 PM. In 1840, daily precipitation readings began. Following the Civil War when no readings were taken, minimum/maximum thermometers were used along with the previous multi reading procedure. In the 1870s and 1880s there were many months with missing data and during the period from 1900 through 1904 no readings were taken. For climate change studies the data for these periods have had to be extrapolated from other nearby stations. The present daily readings of high, low and time of observation temperatures are taken by maintenance employees of the academy. In 1946 the station at an elevation of 167 feet was relocated westward to its present site at an elevation of 360 feet above sea level. There has been only one small but significant location change in the early 1950s when the thermometer shelter was moved from a concrete floor yard enclosure to a grassy lawn. These 20th century site changes have certainly affected monthly averages, but the exact amounts remain uncertain.

ALBANY CITY / AIRPORT

While daily temperature readings at the airport (elevation 275 feet above sea level) began in 1938, this station at its Albany City location has a history dating back to 1813, which makes it one of the five oldest active stations in the United States. Breaks in the record, however did occur from 1815 -1819, and 1861, plus months missing in the mid 1800s. Until 1874 daily temperature readings were taken three times a day from one thermometer with a formula used to determine the mean for each day. The 19th century means should be accepted with caution. Within Albany City from 1826 to 1935 there were ten relocations of the thermometers at elevations ranging from about 100 to 200 feet above sea level with unspecified distances between sites. Published airport daily readings began in 1943. Concurrent readings from the city and airport from1938 to 1970 together with extrapolated earlier data from nearby stations have enabled a continuous temperature record to be obtained for nearly two centuries.

SARATOGA SPRINGS

Temperature observations by volunteers began in 1890 and continued to 1903 with year long breaks in 1893, 1894, 1896, and 1897. Readings to the present began again in 1941 at the Fur Animal Experimental Station 3.5 miles northwest of the post office at an elevation of 550 feet above sea level. Readings continued through 1951 and were not resumed till 1956. In 1991, the station was relocated 4 miles southwest of the post office at an elevation of 310 feet. Readings there are being taken by employees of the New York State Department of Environmental Conservation.

SLIDE MOUNTAIN

This station in Ulster County is located near the summit of the highest elevation in the Catskill Mountains. Its location 4.2 miles northwest of the Oliverea post office at an elevation of 2650 feet above sea level makes it the highest station in New York State. While precipitation monitoring began in 1917, daily temperature readings started in 1940 and are taken by employees of the New York City Department of Water Supply. Since 1992 readings have taken by Timothy Hinkley.

MILLBROOK

Daily readings at the Millbrook School began in 1942, at an elevation of 796 feet, taken by Xavier Prum till his death in 1951. He was succeeded by Neale Howard. School readings terminated in 1999. Concurrent readings began in 1989 at the nearby Institute of Ecosystem Studies, and have been recorded there by Vicki Kelly. These remain unpublished in the New York Climatological Data official National Weather Service monthly summary.

WESTCHESTER AIRPORT 1949 – 2012;
WHITE PLAINS 1873 - 1892

While 20th century daily records began in 1949, three decades of dailytemperature and precipitation were taken in the late 19th century. Daily measurements are presently taken by airport employees and by an automatic ASOS system. The station is at an elevation of 400 feet above sea level on a flat area not typical of the surrounding region. Until 1980 the thermometers were sited between a black topped roadway and the edge of the airfield. Since then the instruments are housed between two airfield runways. In all likelihood winter temperatures are cooler than what is typical for the area and summer temperatures are warmer than what would be representative. Whether these average out to give a true average annual is an open question.

SITE LOCATION AND DATA LEGITMACY

In any evaluation of weather data from a given station, the question that arises, as to whether the data collected truly represents the area in which it is located. A satisfactory answer to this question depends on the criteria used and just what is measured. The criterion for rainfall legitimacy is different from that of snowfall and that of temperature. Unfortunately the placement of a weather station is rarely made on the basis of its usefulness in representing a specific region.

If the station is sited at a public institution it may be placed on the roof of a building. While this may be satisfactory for rainfall measurements, it is not for snowfall or temperature. If the station is sited in a courtyard it may be shielded on one side by a wall or building, in which case all three weather phenomena would be affected and not representative of the surrounding area. Aside from the poor location examples mentioned, to what degree does elevation or proximity to residential or other urbanized regions affect climate? Numerous studies have been done and a formula has been determined in which a temperature correction factor is employed based on the size of the population nearest the weather station. These corrections have been made for the long term stations of the U.S. Historical Climatology Network.

The change of the location of the station is another obvious factor affecting all three measured parameters. Elevation and topography changes along with a relocation from an urban setting to a more rural one can also impact measurements. A classic example is the Albany station. While it is the oldest in New York State it has had numerous site changes in its nearly 200 year history. Adjustments and estimations have had to be made in order to arrive at continuous useable climate history .

TIME OF OBSERVATION ADJUSTMENTS

Ideally daily temperature readings should be taken at midnight as the high and low temperature would then be recorded for the same calendar day. However at nearly all weather stations, for the convenience of the observer, temperatures are recorded in the morning or late afternoon. This results in a bias so that the monthly averages for stations that take their readings in the morning are generally lower than if they were taken at midnight. Stations taking their readings in the late afternoon have their monthly averages biased on the high side. Depending on the month and hour of observation time these differences can be as great as two degrees. In order to correct for these biases a monthly correction factor has to be applied, which is usually different for each month and hour of observation.

INDEX

INDEX

INDEX

TOWNS & LOCATIONS:

www.ingramcontent.com/pod-product-compliance
Lightning Source LLC
Chambersburg PA
CBHW081504170526
45166CB00008B/2546